Advances in Intelligent Systems and Computing

Volume 805

Series editor

Janusz Kacprzyk, Polish Academy of Sciences, Warsaw, Poland
e-mail: kacprzyk@ibspan.waw.pl

The series "Advances in Intelligent Systems and Computing" contains publications on theory, applications, and design methods of Intelligent Systems and Intelligent Computing. Virtually all disciplines such as engineering, natural sciences, computer and information science, ICT, economics, business, e-commerce, environment, healthcare, life science are covered. The list of topics spans all the areas of modern intelligent systems and computing such as: computational intelligence, soft computing including neural networks, fuzzy systems, evolutionary computing and the fusion of these paradigms, social intelligence, ambient intelligence, computational neuroscience, artificial life, virtual worlds and society, cognitive science and systems, Perception and Vision, DNA and immune based systems, self-organizing and adaptive systems, e-Learning and teaching, human-centered and human-centric computing, recommender systems, intelligent control, robotics and mechatronics including human-machine teaming, knowledge-based paradigms, learning paradigms, machine ethics, intelligent data analysis, knowledge management, intelligent agents, intelligent decision making and support, intelligent network security, trust management, interactive entertainment, Web intelligence and multimedia.

The publications within "Advances in Intelligent Systems and Computing" are primarily proceedings of important conferences, symposia and congresses. They cover significant recent developments in the field, both of a foundational and applicable character. An important characteristic feature of the series is the short publication time and world-wide distribution. This permits a rapid and broad dissemination of research results.

More information about this series at http://www.springer.com/series/11156

Edgardo Bucciarelli · Shu-Heng Chen
Juan Manuel Corchado
Editors

Decision Economics. Designs, Models, and Techniques for Boundedly Rational Decisions

 Springer

Editors
Edgardo Bucciarelli
Department of Philosophical, Pedagogical
and Economic-Quantitative Sciences,
Section of Economics and Quantitative
Methods
University of Chieti-Pescara
Pescara, Italy

Juan Manuel Corchado
Departamento de Informática y Automática
Universidad de Salamanca
Salamanca, Spain

Shu-Heng Chen
AI-ECON Research Center - Department
of Economics
National Chengchi University
Taipei, Taiwan

ISSN 2194-5357   ISSN 2194-5365   (electronic)
Advances in Intelligent Systems and Computing
ISBN 978-3-319-99697-4   ISBN 978-3-319-99698-1   (eBook)
https://doi.org/10.1007/978-3-319-99698-1

Library of Congress Control Number: 2018953023

This Springer imprint is published by the registered company Springer Nature Switzerland AG
The registered company address is: Gewerbestrasse 11, 6330 Cham, Switzerland

The photo of Herbert A. Simon was found in the public domain.

To Herbert A. Simon.
(The Editors)

Preface

Decision Economics. Designs, Models and Techniques for Boundedly Rational Decisions

Decision economics is a growing field of research which has been given much attention by several scholars in recent decades. The special session on Decision Economics (DECON) is a multidisciplinary international forum dedicated to advancing knowledge and improving practice in the areas of economics, business, computer science, cognitive sciences, quantitative methods, and related disciplines. To pursue this mission, DECON has facilitated the development and dissemination of a curriculum programme package concerning the diverse disciplines of the decision sciences by seeking to reconcile theory and practice in the tradition of Herbert A. Simon's interdisciplinary heritage. Indeed, Herbert A. Simon was—and still is—one of the most influential founding father of the multidisciplinary fields of decision sciences: his contributions range across administrative behaviour and organisation theory, management science and operation research, behavioural decision theory, cognitive psychology, and artificial intelligence.

In 2018, DECON reported a third consecutive year of record-breaking activity based on theoretical, empirical, and methodological investigations of socio-economic decisions made by several economic agents in a complex market economy, not at all according to standard "rational" economic principles, but especially in light of behavioural and cognitive factors and bounded rationality. Such investigations have focused methodologically on quantitative approaches, qualitative methods, or have taken the form of insightful logic and flowing argumentation as well as reviews and commentaries on best practices in social science research. Furthermore, the special session has particularly dealt with analytics as an emerging synthesis of sophisticated methodology and large data systems used to guide economic decision-making in an increasingly complex business environment.

The recent negative economic and financial events, which have hit the world economies over approximately the last decade (2007–2017), call for new and innovative studies mainly in economics, business and finance, involving different research fields such as economics, psychology or strictly related social sciences, leading source for original research on the interplay of those fields within computer science and artificial intelligence. The editors of this book, chairs of DECON, strongly believe that this is a moral duty, as well as a scientific duty, for prudent and wise economists aware of the complexity of the real world in which they live and work in the third millennium. Certainly, the combination of economics and decision sciences is a field of studies which proves to be useful and, thus, should be fostered to academicians and practitioners interested in the application of quantitative, behavioural and cognitive methods to the problems of society. The special session has focused on interdisciplinary approaches to the study of economic analysis and policies within several major areas, as shown below among others:

- Experimental Research in Economics; Behavioural Game Theory; Cognitive Economics; Interrelations of Economics and Psychology with AI;
- Complexity of Behavioural Decision Processes; Microfoundations and Micro–Macro Relationships; Organisational Decision-Making;
- Computation and Computability in Economics; Decision Theory and the Economics of Uncertainty; Algorithmic Social Sciences Research;
- Decision Support Systems and Business Decisions; Business Intelligence Analytics and Decision-Making; Big Data, Data Mining and Robotics;
- Managerial Decision-Making; Complex Business Environments; Information Technology and Operational Decision Sciences.

This book presents a collection of selected peer-reviewed papers presented at DECON 2018 and discussed decision economics from a wide spectrum of methodological issues and applications. The content of each chapter is described briefly below.

Chapter 1. *"Auto regressive integrated moving average modeling and support vector machine classification of financial time series"* by Alexander Kocian and Stefano Chessa. In this chapter, with cloud computing offering organisations a level of scalability and power, the authors are at a point where machine learning is set to support human financial analysts in FOReign EXchange (FOREX) markets. Trading accuracy of current robots, however, is still hard limited. They deal with the derivation of a one-step predictor for a single FOREX pair time-series. In contrast to many other approaches, the authors' predictor is based on a theoretical framework. The historical price actions are modelled as autoregressive integrated moving average (ARIMA) random process, using maximum likelihood fitting. The minimum akaike information criterion estimation (MAICE) yields the order of the process. A support vector machine (SVM), whose feature space is spanned by historical price actions, yields the one-step-ahead class label UP or DOWN. Backtesting results on the EURUSD pair on different time frames indicate that their predictor is capable of achieving high instantaneous profit but on long-term average, profitable when the risk-to-reward ratio per trade is around 1:1.2. The authors' result is in line with related studies.

Chapter 2. *"Do information quantity and transmission make a difference to the stable contrarian?"* by Hung-Wen Lin, Jing-Bo Huang, Kun-Ben Lin and Shu-Heng Chen. In this chapter, the authors study how financial transparency and media coverage work in the Chinese stock markets. In this work, transparency means information quantity, while media means information transmission. The market has negative momentum profits no matter how transparency or media coverage changes, which suggests that transparency—or media coverage—does not work individually in China. High transparency and high media coverage make significantly positive momentum profits, whereas low transparency and low media coverage make significantly negative momentum profits. These outcomes show that transparency and media coverage work jointly in China. The authors' findings imply that information quantity and transmission are both crucial in China.

Chapter 3. *"The logistic map: an AI tool for economists investigating complexity and suggesting policy decisions"* by Carmen Pagliari and Nicola Mattoscio. In this chapter, the authors give an original interpretation of the logistic map popularised by the biologist Robert May in 1976. This map is potentially a powerful AI tool based on a deterministic methodology having a double opportunity to be applied in economics. On the one hand, indeed, the first application concerns the investigation of a certain intrinsic complexity of real economic phenomena characterised by endogenous nonlinear dynamics. Closely related to a normative research, on the other hand, the second application helps determine results useful for suggesting policy decisions as a contribution to avoiding chaos and unpredictability within real economic systems. In the first application, therefore, the logistic map can be used as an AI tool for forecasting and anticipating the unknown (for previsions of bifurcations, cycles and chaos), while in the second one, it can be considered as an AI tool for policymakers in order to deduce the analytical conditions that aid the economic system in being sufficiently far away from chaos and uncontrollability.

Chapter 4. *"Optimal noise manipulation in asymmetric tournament"* by Zhiqiang Dong and Zijun Luo. In this chapter, the authors fill a gap in the literature of asymmetric tournament by allowing the principal to optimally alter noise in relative performance evaluation, such that the observed performance of each agent is less or more dependent of ability and effort. The authors show that there exists an optimal noise level from the principal's standpoint of expected profit maximisation. It is shown that this optimal noise level is higher than what would induce the highest efforts from the two agents.

Chapter 5. "Decision-making process underlying travel behavior and its incorporation in applied travel models" by Peter Vovsha. In this chapter, the author provides a broad overview of the state of the art and practice in travel modelling in its relation to individual travel behaviour. The work describes how different travel decision-making paradigms led to different generations of applied travel models in practice—from aggregate models to disaggregate trip-base models, then to tour-based models, then to activity-based models and finally to agent-based models. The chapter shows how these different modelling approaches can be effectively generalised in one framework where different model structures correspond to different basic assumptions on the decision-making process. The author focus on three key underlying behavioural aspects: (i) how different dimensions of travel and associated individual choices are sequenced and integrated; (ii) how the real-world constraints on different travel dimensions are represented; and (iii) what are the behavioural factors and associated mathematical and statistical models applied for modelling each decision-making step. The work analyses the main challenges associated with understanding and modelling travel behaviour and outlines avenues for future research.

Chapter 6. *"Formalisation of situated dependent-type theory with underspecified assessments"* by Roussanka Loukanova. In this chapter, the author introduces a formal language of situated dependent-type theory by extending its potentials for structured data integrated with quantitative assessments. The language has terms for situated information which is partial and underspecified. The enriched formal language provides integration of a situated dependent-type theory with statistical and other approaches to machine learning techniques.

Chapter 7. *"Scholarship, admission and application of a postgraduate program"* by Yehui Lao, Zhiqiang Dong and Xinyuan Yang. In this chapter, the authors aim to construct a game of admission and application behaviour of a postgraduate programme. They attempt to expand the decision of graduate school from one party model to two party model. The authors suggest that the interval of postgraduate programme's scholarship determines the decision made by applicants with different capacity and family background. Furthermore, graduate schools will try to use scholarship as a tool to select students.

Chapter 8. *"Subgroup optimal decisions in cost-effectiveness analysis"* by Elias Moreno, Francisco-José Vázquez-Polo, Miguel-Angel Negrín and María Martel-Escobar. In this chapter, the authors deal with cost-effectiveness analysis (CEA) of medical treatments. In this framework, the optimal treatment is chosen using a statistical model of the cost and effectiveness of the treatments and data from patients under the treatments. Sometimes, however, these data also include values of certain deterministic covariates of the patients with usually have valuable clinical information that would be incorporated into the statistical treatment selection procedure. In this respect, the authors discuss the usual statistical models to undertake this task and the main statistical problems it involves. They present a Bayesian variable selection procedure and find optimal treatments for subgroups defined by selected covariates.

Chapter 9. *"A statistical tool as a decision support in enterprise financial crisis"* by Francesco De Luca, Stefania Fensore and Enrica Meschieri. In this chapter, the authors focus on the recent reform of Italian Insolvency Law (IIL) which has introduced new instruments aimed to restore companies bearing financial distresses and potentially incurring bankruptcy proceedings. In particular, the Article 182-bis restructuring agreements have been introduced by the Italian Civil Code to manage, among others, these distresses and potential proceedings. Therefore, the authors' objective is to underline the ability of seven specific accounting ratios and coefficients to help predict the status of financial distress of companies. The authors introduce a new formula that they call M-Index indicator and then provide an empirical analysis through a sample of Italian listed companies collected from Borsa Italiana (Italian Stock Exchange) in the period 2003–2012. The results of the empirical analysis performed by the authors validate the predictive accuracy power of their indicator.

Chapter 10. *"A mediation model of absorptive and innovative capacities: The case of Spanish family businesses"* by Felipe Hernández-Perlines and Wenkai Xu. In this paper, the authors analyse the mediating effect of innovation capacity on the influence of absorptive capacity in the performance of family businesses. For the analysis of results, the use of a second-generation structural equation method is proposed (PLS-SEM) using smartPLS 3.2.7 computer software, applied to the data coming from 218 Spanish family businesses. The main contribution of this work is given by the fact that the performance of family businesses is determined by the absorptive capacity (absorptive capacity is able to explain approximately 36% of the performance variability of family businesses). The second relevant contribution of this work is that the influence of the absorptive capacity on the performance of family businesses is strengthened by the effect of innovation capacity, explaining around 40% of the variability. The third contribution is that the absorptive capacity is a precedent for innovation capacity, able to explain about 50% of its variability.

Chapter 11. *"The mathematics of interdependence for superordinate decision-making with teams"* by William Lawless. In this chapter, the author reviews the function of decision-making as a human process in the field affected by interdependence (additive and destructive social interference). The scope of this review is first to define and describe why interdependence is difficult to grasp intellectually, but much easier intuitively in social contexts (bistability, convergence to incompleteness, non-factorable social states); second to describe the research accomplishments and applications to hybrid teams (arbitrary combinations of humans, machines and robots); and third to advance the research by beginning to incorporate the value of intelligence for teams as they strive to achieve a team's superordinate goals (e.g. in the tradeoffs between a team's structure and its effort to achieve its mission with maximum entropy production, MEP). The author discussed prior results, future research plans and draw conclusions for the development of theory.

Chapter 12. *"Towards a natural experiment leveraging big data to analyse and predict users"* behavioural patterns within an online consumption setting" by Raffaele Dell'Aversana and Edgardo Bucciarelli. In this chapter, the authors develop a model for multi-criteria evaluation of big data within organisations concerned with the impact of an ad exposure on online consumption behaviour. The model has been structured to help organisations make decisions in order to improve the business knowledge and understanding on big data and, specifically, heterogeneous big data. The model accommodates a multilevel structure of data with a modular system that can be used both to automatically analyse data and to produce helpful insights for decision-making. This modular system and its modules, indeed, implement artificial intelligent algorithms such as neural networks and genetic algorithms. To develop the model, therefore, a prototype has been built by the authors as proof-of-concept using a marketing automation software that collects data from several sources (public social and editorial media content) and stores them into a large database so as the data can be analysed and used to help implement business model innovations. In this regard, the authors are conducting a natural experiment which has yet to be completed in order to show that the model can provide useful insights as well as hints to help decision-makers take further account of the most 'satisficing' decisions among alternative courses of action.

Chapter 13. *"Google trends and cognitive finance: Lessons gained from the Taiwan stock market"* by Pei-Hsuan Shen, Shu-Heng Chen and Tina Yu. In this chapter, the authors investigate the relationship between Google Trends Search Volume Index (SVI) and the average returns of Taiwan Stock Exchange Capitalization Weighted Stock Index (TAIEX). In particular, the authors used the aggregate SVI searched by a company's abbreviated name and by its ticker symbol to conduct our research. The results are very different. While the aggregate SVI of abbreviated names is significantly and positively correlated to the average returns of TAIEX, the aggregate SVI of ticker symbols is not. This gives strong evidence that investors in the Taiwan stock market normally use abbreviated names, not ticker symbols, to conduct Google search for stock information. Additionally, the authors found the aggregate SVI of small–cap companies has a higher degree of impact on the TAIEX average returns than that of the mid–cap and large–cap companies. Finally, the authors found the aggregate SVI with an increasing trend also has a stronger positive influence on the TAIEX average returns than that of the overall aggregate SVI, while the aggregate SVI with a decreasing trend has no influence on the TAIEX average returns. This supports the attention hypothesis of Terrance Odean in that the increased investors attention, which is measured by the Google SVI, is a sign of their buying intention, hence caused the stock prices to increase while decreased investors attention is not connected to their selling intention or the decrease of stock prices.

Chapter 14. *"Research on the evaluation of scientists based on weighted h-index"* by Guo-He Feng and Xing-Qing Mo. In this chapter, the authors proposed two weighted h-index models which are named hw-index and hw_t-index and then selected 30 active Chinese scholars in Llbrary and information science field for empirical analysis. Revealing highly cited papers and considering the contribution of scientists' whole papers, hw-index not only weakens the influence of self-citation on the results, but also makes it easy to distinguish scientists' contributions. The hw_t-index focuses on the research output of scientists in recent years, also considering their past contributions. Therefore, in the short-term evaluation, the hw_t-index is more reasonable for young scientists and scholars who made a great contribution during the past years. Potential scholars can be identified by the way of comparing hw-index with hw_t-index.

Chapter 15. *"Decision analysis based on artificial neural network for feeding an industrial refrigeration system through the use of photovoltaic energy"* by Fabio Porreca. In this chapter, the author deals with the evaluation of the energy availability from renewable sources in the industrial processes. The subject matter of the research is at the basis of many studies concerning engineering applications. The non-programmable nature of many of the renewable sources often leads to consider them as a simple support and not as a primary source of supply. With this in mind, the author tries to exploit the forecasting abilities of the neural networks in order to create scenarios applicable in different high-energy-consuming industrial contexts which reckon the optimisation of the energy consumption as the new objective of the so-called green business.

Chapter 16. *"Exit, voice and loyalty in consumers"* online-posting behaviour: An empirical analysis of reviews and ratings found on Amazon.com" by Tony Ernesto Persico, Giovanna Sedda and Assia Liberatore. In this chapter, the authors aim to describe e-commerce consumers' behaviour by analysing the distribution of online reviews and ratings. Different from previous studies focused on the positivity and negativity of ratings, this work analyses the ratings distribution through a tensor-based approach. This approach allows the authors to observe a new range of information related to distributions' features that they describe through the "Exit, Voice and Loyalty" scheme. In addition, the authors seek a distribution function capable of capturing these features, and they aim to over-perform the synthesis provided by using a polynomial regression model. For this reason, the authors introduce an ad hoc beta-type modified function to create a proxy of collected data. Finally, the authors found a tri-modal distribution (S-modal) as a relevant component of the J-shaped distributions referred in the literature.

Chapter 17. "Effective land-use and public regional planning in the mining industry: The case of Abruzzo" by Francesco De Luca, Stefania Fensore and Enrica Meschieri. In this chapter, the authors deal with issues concerning land use patterns, public planning and extractive industry. More specifically, the authors aim to help describe the socio-economic variables most affected by quarry extraction processes by referring to the case of Abruzzo (region of Central Italy). The authors, moreover, introduce a model for quantification of the quarry material requirements expressed by the economic operators of the same territory with a time horizon of 2020. To this end, they suggest the use of several economic and statistical indicators, such as public investment on infrastructures, GDP growth, social housing policies and private building permits, in order to optimise the predicting power of the model as the indicators represent reliable proxies of the demand of raw materials, with respect to the need to limit the impact on the natural environment.

Chapter 18. *"Do ICTs matter for Italy?"* by Daniela Cialfi and Emiliano Colantonio. In this chapter, the authors investigate the relationship between ICTs and social capital through the study of the relative disparities among Italian regions. This work provides an operational definition of the concepts of ICT and social capital and presents consistent evidence on the geography of this relationship in Italy. The statistical and geographical analysis, based on nonlinear clustering with self-organising map (SOM) neural networks, are performed to analyse the performance of Italian regions in the period 2006–2013. The results show the isolation of Southern Italian regions. In particular, the authors found that ICTs may not promote social capital; that is, ICTs could not play a decisive role in creating and developing social capital. These results prompt the formulation of new policies for Italian regions.

Chapter 19. "Relationship of Weak Modularity and Intellectual Property Rights for Software Products" by Stefan Kambiz Behfar and Qumars Behfar. In this chapter, the authors focus on the impact of modularity on intellectual property rights, referring to modularity of underlying products to capture value within firms. In particular, the authors bring together the theory of software modularity from computer science and intellectual property (IP) rights from management literature to address the issue of value appropriation for IP rights within software products. The work defines the term of intellectual property associated with software products or platforms as opposed to the term of intellectual property used within firms serving as a source of economic rents. Initially, the work discusses the concepts behind usage of modularity as a means to protect IP rights and explain differences of organisation and product modularity, while rendering calculation for probability of imitation for weak modular systems. Then, the work investigates the threat of imitation. The main contribution of this paper is to provide a systematic analysis of value appropriation in weak modular systems by introducing a relationship between probability of imitation and module interdependency.

Last but not least, once again, this year's special session on Decision Economics would not have been possible without the advice and support of many scholars, particularly those belonging to the Programme Committee. Among these scholars, furthermore, special thanks are due to Sara Rodríguez González and Fernando De la Prieta for their dedicated guidance as well as their time, generosity and comments. The programme committee members and English language mentors have helped ensure the quality of the contributions with their extensive and continuous feedback to most of the authors and the editors, too.

Edgardo Buciarelli
Shu-Heng Chen
Juan Manuel Corchado

Organisation

Organisation of the Special Session on Decision Economics 2018

Chairs

Edgardo Bucciarelli   University of Chieti-Pescara, Italy
Shu-Heng Chen   National Chengchi University, Taipei, Taiwan

International Programme Committee

Federica Alberti   University of Portsmouth, UK
José Carlos R. Alcantud   University of Salamanca, Spain
Barry Cooper (Deceased)   University of Leeds, UK
Sameeksha Desai   Indiana University, Bloomington, USA
Zhiqiang Dong   South China Normal University, China
Felix Freitag   Universitat Politècnica de Catalunya, Spain
Jakob Kapeller   Johannes Kepler University of Linz, Austria
Amin M. Khan   IST, University of Lisbon, Portugal
Alan Kirman   Aix-Marseille Université, France
Alexander Kocian   University of Pisa, Italy
Nadine Levratto   Université Paris Ouest Nanterre La Défense, France
Nicola Mattoscio   University of Chieti-Pescara, Italy
Elías Moreno   University of Granada, Spain
Giulio Occhini   Italian Association for Informatics and Automatic Calculation, Milan, Italy
Luigi Orsenigo   IUSS, University of Pavia, Italy

Lionel Page	Queensland University of Technology, Brisbane, Australia
Enrico Rubaltelli	University of Padua, Italy
Anwar Shaikh	The New School for Social Research, New York, USA
Pietro Terna	University of Turin, Italy
Katsunori Yamada	Osaka University, Japan
Ragupathy Venkatachalam	Goldsmiths, University of London, UK
Stefano Zambelli	University of Trento, Italy

Contents

About the Editors

Edgardo Bucciarelli is an Italian economist who holds the position of research professor of economics at the University of Chieti-Pescara (Italy), where he earned his PhD in economics (cv SECS/P01). His main research interests lie in the area of complexity and market dynamics; decision theory, design research, experimental microeconomics, classical behavioural economics and economic methodology. His main scientific articles appeared, among others, in the Journal of Economic Behavior and Organization, Journal of Post Keynesian Economics, Metroeconomica, Applied Economics, Computational Economics and other international journals. Several key contributions appeared in chapters of book in Physica-Verlag and Springer Lecture Notes in Economics and Mathematical Systems. At present, he teaches experimental economics, cognitive economics and finance and economic methodology at the University of Chieti-Pescara. He is one of the directors of the Research Centre for Evaluation and Socio-Economic Development and the co-founder of the academic spin-off company "Economics Education Services". He is the co-founder, organising chair, programme committee chair in a number of international conferences.

Shu-Heng Chen is a Taiwanese economist. He earned his PhD in economics at the University of California (UCLA, Los Angeles, USA) in 1992. Currently, he is a distinguished professor of economics in the Department of Economics and also the dean of the Office of International Cooperation at the National Chengchi University (Taipei, Taiwan). Furthermore, he is the founder and director of the AI-ECON Research Center at the College of Social Sciences of the National Chengchi University and the coordinator of the Laboratory of Experimental Economics in the same University. He is unanimously considered one of the most influential and pioneer scholars in the world in the field of applied research known as computational economics. His scientific contributions were directed to the affirmation of the computational approach aimed to the interpretation of the theoretical issues and applied economic problems still today unresolved, from a perspective more connected to reality and therefore different from the dominant neoclassical paradigm. In particular, his most decisive contributions are aimed to the approach based on models with heterogeneous agents and the genetic programming in the socio-economic studies. His work as a scholar is interdisciplinary and focused since the beginning on methodologies related to the bounded rationality and Herbert A. Simon's contributions. Shu-Heng Chen holds the position of editor of prestigious international economic journals and is author of more than 150 publications including scientific articles, monographs and book chapters.

Juan Manuel Corchado is a full professor with Chair at the University of Salamanca. He was the vice president for Research and Technology Transfer from December 2013 to December 2017 and the director of the Science Park of the University of Salamanca, director of the Doctoral School of the University until December 2107, and also, he has been elected twice as the dean of the Faculty of Science at the University of Salamanca. In addition to a PhD in computer sciences from the University of Salamanca, he holds a PhD in artificial intelligence from the University of the West of Scotland. Juan Manuel Corchado is a visiting professor at Osaka Institute of Technology since January 2015, visiting professor at the University Teknologi Malaysia since January 2017 and a member of the Advisory group on Online Terrorist Propaganda of the European Counter Terrorism Centre (EUROPOL). Corchado is the director of the BISITE (Bioinformatics, Intelligent Systems and Educational Technology) Research Group, which he created in the year 2000, president of the IEEE Systems, Man and Cybernetics Spanish Chapter, academic director of the Institute of Digital Art and Animation of the University of Salamanca. He also oversees the master's programmes in digital animation, security, mobile technology, community management and management for TIC enterprises at the University of Salamanca. Corchado is also the editor and editor-in-chief of specialised journals like ADCAIJ (Advances in Distributed Computing and Artificial Intelligence Journal), IJDCA (International Journal of Digital Contents and Applications) and OJCST (Oriental Journal of Computer Science and Technology).

Auto Regressive Integrated Moving Average Modeling and Support Vector Machine Classification of Financial Time Series

Alexander Kocian[(✉)] and Stefano Chessa

Department of Computer Science, University of Pisa, 56123 Pisa, Italy
kocian@di.unipi.it

Abstract. With cloud computing offering organizations a level of scalability and power, we are finally at a point where machine learning is set to support human financial analysts in FOReign EXchange (FOREX) markets. Trading accuracy of current robots, however, is still hard limited.

This paper deals with the derivation of a one-step predictor for a single FOREX pair time-series. In contrast to many other approaches, our predictor is based on a theoretical framework. The historical price actions are modeled as Auto Regressive Integrated Moving Average (ARIMA) random process, using maximum likelihood fitting. The Minimum Akaike Information Criterion Estimation (MAICE) yields the order of the process. A Support Vector Machine (SVM), whose feature space is spanned by historical price actions, yields the one-step ahead class label UP or DOWN.

Backtesting results on the EURUSD pair on different time frames indicates that our predictor is capable of achieving high instantaneous profit but on long-term average, is only profitable when the the risk-to-reward ratio per trade is around 1:1.2. The result is inline with related studies.

Keywords: Machine learning · Support Vector Machine · ARIMA FOREX · Probability theory

1 Introduction

Decision theory is an interdisciplinary approach to study the reasoning of an underlying agent's choice. Decision theory, closely related to game theory, can be classified in (i) normative decision theory, focusing on how perfectly rational agents should make their decisions absent all constraints such as time, risk, etc.; (ii) descriptive decision theory, analyzing how (irrational) agents actually make decisions; and (iii) prescriptive decision theory, focusing on a feasible procedure in line with the normative theory.

© Springer Nature Switzerland AG 2019
E. Bucciarelli et al. (Eds.): DCAI 2018, AISC 805, pp. 1–8, 2019.
https://doi.org/10.1007/978-3-319-99698-1_1

Following the latter approach, we focus on the financial market. Financial decisions must often take into account future events, such as FOReign EXchange (FOREX), stocks or portfolios. Economic theory is used to evaluate how time, risk, opportunity costs and information can stimulate a decision.

The FOREX market is the largest financial market in the world with an average spot volume of around US\$ 5,1 trillion in 2016, according to the Bank of International Settlements (BIS). The EUR and the USD had the biggest shares with US\$ 4,44 trillion and US\$ 1,59 trillion average transactions a day, respectively. Unlike stocks, currency trading is not controlled by any central governing body, there are no clearing houses to guarantee the trades. All members trade with each other based on credit agreements.

To optimize profit, research interest on trading robots has accelerated during the last years. Trading system might be classified in model-based and data-mining based. The first starts with a model of the market anomaly and builds a trading strategy on top of it. The latter approach, in contrast, scans price curves for predictive patterns. In this article we concentrate on the latter method.

Popular data-mining approaches are

- Bayes methods: provide us with a formal way to incorporate the prior information on the market data. It fits perfectly with sequential learning and decision making and it directly leads to exact small sample results.
- Linear regression: Assuming a linear relation ship between input and output, the model combines a specific set of input values to optimize some metric such as the least square error. Support vector machines (SVM), which are considered in this contribution, interpret the inputs as coordinates of a N-dim. feature space, to compute a hyperplane separating the output samples larger zero from those less zero [5,8];
- Non-linear regression: neural-networks learn by determining the coefficients that minimize the error between sample prediction and sample target. The weight coefficients are optimized during backwards propagation of the estimation error [3]. More advanced non-linear learning methods include the Restricted Boltzmann Machine (RBM) and the Sparse Autoencoder (SAE);
- Price actions: Based on open, high, low and close of "candles" in the price chart, one hopes to find patterns that predicts the class of the next pattern;

In this contribution, we use data-mining to represent the features price range and price change as ARIMA stochastic processes. The ARIMA model is than used to derive a one-step ahead predictor. At the same time, this features span the the feature space of an SVM, classifying the predicted price range and price change as UP or DOWN market signal.

2   Support Vector Machines

Let us first review the SVM [2]. Suppose we have a set of T training data points $\mathcal{D} = \{(\boldsymbol{x}_i, y_i)\}$, $i = 1, \ldots, T$, where each member is a pair of a input vector

$x_i \in \mathbb{R}^N$ and a class label $y_i \in \{-1, +1\}$. The classes -1 and $+1$ could represent down and up market movements, respectively. The question is how to divide up the decision boundaries.

Linear Discriminant. We start with the scenario, where the data points are linearly separable, as is illustrated in Fig. 1 for $N = 2$ input parameters.

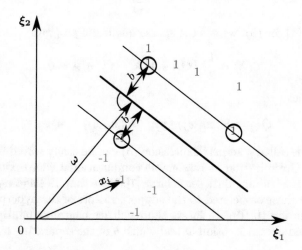

Fig. 1. Series of market observations (down:-1/up:$+1$) on $N = 2$-dimensional feature space. The linear separable case.

The task is to find the separating hyperplane, characterized by the distance b of the closest example to the separating hyperplane and the weight vector $\boldsymbol{\omega}$ which is perpendicular to the hyperplane. From Fig. 1, the classifier is given by the linear discriminant

$$f(\boldsymbol{x}) = \text{sign}\{\boldsymbol{\omega}^\mathsf{T}\boldsymbol{x} + b\} \tag{1}$$

for any $\boldsymbol{x}$ in the feature space. For the sake of convenience, we scale (1) with the class labels y_i. Following this approach, it follows for each data point on the boundary that

$$y_i(\boldsymbol{\omega}^\mathsf{T}\boldsymbol{x}_i + b) - 1 = 0 \tag{2}$$

such that the margin $2b = 2/|\boldsymbol{\omega}|$ is maximized or equivalently, $\frac{1}{2}\boldsymbol{\omega}^\mathsf{T}\boldsymbol{\omega}$ is minimized. From (2) and above boundary condition, the Lagrange function $\mathcal{L}$ yields

$$\mathcal{L}(\boldsymbol{\lambda}, b, \boldsymbol{\omega}) = \frac{1}{2}\boldsymbol{\omega}^\mathsf{T}\boldsymbol{\omega} - \sum_i \lambda_i \left(y_i(\boldsymbol{\omega}^\mathsf{T}\boldsymbol{x}_i + b) - 1\right) \tag{3}$$

where λ_i denote the Lagrange multipliers. The solution to (3) is given by the Karush-Kuhn-Tucker conditions ([4], Chap. 9)

$$\nabla_\omega \mathcal{L}(\boldsymbol{\lambda}, b, \boldsymbol{\omega}) = \boldsymbol{\omega} - \sum_{i=1}^{T} \lambda_i y_i \boldsymbol{x}_i = \mathbf{0},$$
$$\nabla_b \mathcal{L}(\boldsymbol{\lambda}, b, \boldsymbol{\omega}) = - \sum_{i=1}^{T} \lambda_i y_i = 0. \tag{4}$$

Substituting (4) for (3), we get the sparse quadratic program

$$\mathcal{L}(\boldsymbol{\lambda}) = \frac{1}{2}\boldsymbol{\lambda}^\mathsf{T}\boldsymbol{Q}\boldsymbol{\lambda} + \mathbf{1}^\mathsf{T}\boldsymbol{\lambda}, \quad \text{s.t. } \boldsymbol{y}^\mathsf{T}\boldsymbol{\lambda} = \mathbf{0}, \tag{5}$$

with the entries

$$Q_{i,j} = K(\boldsymbol{x}_i, \boldsymbol{x}_j)y_i y_j; \quad K(\boldsymbol{x}_i, \boldsymbol{x}_j) = \boldsymbol{x}_i^\mathsf{T}\boldsymbol{x}_j, \tag{6}$$

where $K(\cdot, \cdot)$ is called a kernel. Expression (5) can be easily solved using Sequential Minimal Optimization at reasonable computational complexity even if the dimension of the training data set is large [7]. Note that all Lagrange multipliers are zero aside those associated to the support vectors of the hyperplane. Having found the solution to (6), we insert the resulting Lagrange multipliers λ_i into (4) and (2) and obtain location $\boldsymbol{\omega}$ and width b of the separating hyperplane.

Non-linear Discriminant. When data is not linearly separable, the linear classifier in (1) cannot fit the data. To solve a non-linear classification problem with a linear classifier, though, we apply a non-linear transform $\Phi(\boldsymbol{x}) : \mathcal{X} \to \mathcal{Z}$ to the input data on the expenses of the space dimension. The mapping is chosen such that the data is linearly separable in the feature space $\mathcal{Z}$. From (1) and (4), the classifier has the general form

$$f(\boldsymbol{x}) = \text{sign}\left\{ \sum_{i=1}^{T} \lambda_i y_i K(\boldsymbol{x}_i, \boldsymbol{x}) + b \right\} \tag{7}$$

with the Kernel function

$$K(\boldsymbol{x}_i, \boldsymbol{x}_j) = \Phi(\boldsymbol{x}_i)^\mathsf{T} \Phi(\boldsymbol{x}_j). \tag{8}$$

Note, that (7) depends on the dimension of the space only implicitly through the inner product in (8). Mercer's condition tells us which kernel functions can be represented as inner product of two vectors. Common kernels, meeting Mercer's condition, are:

- Polynomial kernel: $K(\boldsymbol{x}_i, \boldsymbol{x}_j) = (1 + \boldsymbol{x}_i^\mathsf{T}\boldsymbol{x}_j)^p$ where p is the order of the polynomial,
- Radial kernel: $K(\boldsymbol{x}_i, \boldsymbol{x}_j) = -\gamma\|\boldsymbol{x}_i - \boldsymbol{x}_j\|^2$ with parameter γ,
- Sigmoid: $K(\boldsymbol{x}_i, \boldsymbol{x}_j) = \tanh(a\boldsymbol{x}_i^\mathsf{T}\boldsymbol{x}_j + r)$ with parameters a and r.

3 Auto Regressive Integrated Moving Average

So far, we have trained the SVM classifier on T points. It is of utmost interest for the trader to anticipate the one step ahead class label y_{T+1}. The SVM is capable of doing that if the corresponding input vector x_{T+1} were known. We need to find an estimate $\hat{x}_{T+1}$ of x_{T+1}.

The non-stationary behavior of the time-series makes their prediction cumbersome. Most of the financial series, however, get stationary after d differentiating steps, encouraging us to model the FOREX time series as Auto Regressive Integrated Moving Average (ARIMA) process with parameters (p,d,q), where p and q denote the lag order and the size of the moving average window ([1], Chap. 7). The state-space representation of ARIMA is given by

$$\left(1 - \sum_{k=1}^{p} \alpha_k L^k\right)(1-L)^d v[t] = \left(1 + \sum_{k=1}^{q} \beta_k L^k\right)\epsilon[t] \tag{9}$$

Here, L is the time lag operator, defined as $(Lv)[t] = v[t-1]$, $\forall t \in \mathbb{Z}$, given a sequence $\{v[t]\}$. Hence, the new sequence $(Lv)[t]$ at time t is equal to the original sequence at time $t-1$. Moreover, $\epsilon[t]$ in (9) is a white Gaussian input process with variance σ_ϵ^2. Similarly, $v[t]$ is a Gaussian output process with variance σ_v^2. We need to estimate the parameter vector $\theta = \{\alpha, \beta, \sigma_v^2\}$ and the model order $\omega = \{p, d, q\}$.

We want to compute the log-likelihood function $\Lambda(\theta)$ of the ARIMA(p,d,q) process. For small T, it can be shown ([1], Chap. 7) that

$$\Lambda(\theta) \propto -\frac{T}{2} \log \sigma_v^2 - \frac{\sum_{t=1}^{T} \hat{v}[t] + e^\mathsf{T}\mathrm{Cov}\{e\}^{-1}e}{2\sigma_v^2} \tag{10}$$

with the short-cut $e = [\alpha, \beta]^\mathsf{T}$ and the innovations $\hat{v}[t]$. From (10), the maximum likelihood estimate of θ is then given by

$$\hat{\theta} = \arg\max_{(\theta)} \Lambda(\theta). \tag{11}$$

The optimization problem in (11) can be solved recursively, as outlined in ([1], Chap. 7). The order of the ARIMA process can be estimated by either the Box-Jenkins method [1] or the Minimum Akaike Information Criterion Estimation (MAICE) [6]. Following the latter approach, it can be shown that

$$\mathrm{AIC}(p,d,q) = T \log \hat{\sigma}_v^2 + \frac{2T(p+q+1+\delta)}{T-d} + T \log 2\pi + T, \tag{12}$$

where

$$\delta = \begin{cases} 1 \text{ if } d = 0 \\ 0 \text{ if } d \neq 0 \end{cases}. \tag{13}$$

The tuple $(p^\star, d^\star, q^\star)$ that minimizes the AIC criterion is the optimum [6].

Repeating the procedure in (9) for the remaining $N-1$ features yields an estimate of the one-step ahead data point x_{T+1}.

4   Discussion and Conclusions

The non-linear structure of the SVM makes a theoretical performance analysis cumbersome. To evaluate the performance of our predictor, though, we rely on backtesting in FOREX markets. Specifically, we test the algorithm on the EURUSD currency pair which is currently the most significant in the world.

The feature space of our SVM is spanned by price actions. In particular, we consider $N = 2$ features related to price change $c_i[t]$ and price range $r_i[t]$, defined as

$$c_i[t] \triangleq \frac{\text{Open}[i] - \text{Open}[i-t]}{\text{Open}[i]}, \quad r_i[t] \triangleq \frac{\text{High}[i-t] - \text{Low}[i-t]}{\text{Open}[i]}, \quad (14)$$

$t = 1, \ldots, T$. Each data point in the feature space can then be specified as $\boldsymbol{x}_i = [c_i[0], r_i[0]]^\mathsf{T}$ with class label

$$y_i = \begin{cases} 1 \text{ , Close}[i] \geq \text{Open}[i] \\ -1 \text{ , Close}[i] < \text{Open}[i] \end{cases}. \quad (15)$$

Our SVM-based predictor has been equipped and compared with the following kernel functions:

– SVM-L: the kernel is linear,
– SVM-R: the kernel is radial with $\gamma = 0.5$,
– SVM-S: the kernel is sigmoid with $a = 0.5$ and $r = 0$.

The software application was written and tested in R.

4.1   Methodology

The price information (High,Low,Open,Close) has been taken from the trading platform MetaTrader 4. Daily, four-hour, and one-hour time frames from 2/1/2017-29/12/2017 were subject to the analysis. The data is used to form the price action vectors in (14) with class labels in (15). For each data point i, the price action vectors were fed into (9), to generate an ARIMA stochastic process. Its parameters are the maximum likelihood estimates in (11). The MAICE criterion in (12) selects the model order of the process. The resulting estimate of the current data point $\boldsymbol{x}_i = [c_i[0], r_i[0]]^\mathsf{T}$ is passed to the already trained SVM that outputs a classification according to (7).

4.2   Results

We start with the daily frame. In a first experiment, we evaluated the prediction accuracy A, defined as

$$A(T) \triangleq \frac{\text{\# correct decisions}}{\text{\# total decisions}} 100 \quad [\%], \quad (16)$$

for the year 2017. The result is listed in Table 1 as a function of T. Generally speaking, the SVM-S scheme performs the best among the tested versions. Note that the sigmoid function is a hyperbolic tangent function, implying that data point with small correlation $\boldsymbol{x}_i^\mathsf{T}\boldsymbol{x}_j$, containing high information, have strong impact on the kernel in (8), while all those data points with high correlation, containing low information, are only linearly scaled. Focusing on the SVM-S, it can be seen that the trading accuracy is the best at $T = 60$. When T is further increased, the performance degrades slightly. This is mainly because the SVM over-fits the data at higher risk. For $T = 60$, the trading accuracy of the SVM-S scheme is about 55%, corresponding to a risk-to-reward ratio 1:1.2.

Table 1. Trading accuracy of the proposed predictor for the EURUSD pair on a daily time frame

		Training length T					
		20	30	40	60	80	100
Predictor	SVM-L	49%	49%	50%	52%	49%	49%
	SVM-R	49%	50%	46%	48%	49%	49%
	SVM-P	47%	47%	50%	52%	53%	51%
	SVM-S	**50%**	49%	49%	**55%**	**51%**	**52%**

Figure 2 shows the trading accuracy per month as a function of T for the SVM-S scheme. Aside one drawdown in the month of January, the SVM-S

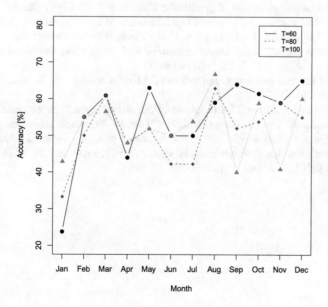

Fig. 2. Accuracy of the SVM-S algorithm with training length T as parameter.

achieves revenue for the year with a risk-to-reward ratio up to 1:1.9. With increasing T, the drawdown eases but the overall trading accuracy drops, as well, mainly due to model over-fitting.

When executed on different time-frames, the considered SVM schemes achieve a very similar trading accuracy. We therefore omitted the presentation here.

4.3 Conclusions

In this contribution, we derived a one-step ahead predictor for the FOREX market based on a theoretical framework. Specifically, the time-series of the features was modeled as ARIMA stochastic process that was passed to a SVM, to predict its class. Backtesting on the most liquid FOREX pair, namely the EURUSD, indicates that the proposed SVM with sigmoid kernel performs best. In this case, the predictor generates profits as long as the risk-to-reward ratio per trade is better than 1:1.2.

To improve trading accuracy, one could tune the parameters of the sigmoid kernel which is subject to future research.

References

1. Box, G.E.P., Jenkins, G.: Time Series Analysis, Forecasting and Control. Holden-Day, Inc., San Francisco (1990)
2. Cortes, C., Vapnik, V.: Support-vector networks. Mach. Learn. **20**, 273–297 (1995)
3. Dixon, M.F., Klabjan, D., Bang, J.: Classification-based financial markets prediction using deep neural networks. Algorithmic Financ. **6**(3–4), 67–77 (2016)
4. Fletcher, R.: Practical Methods of Optimization. Wiley, Chichester (1987)
5. Kara, Y., Boyacioglu, M.A., Baykan, Ö.K.: Predicting firection of stock price index movement using artificial neural networks and support vector machines. Int. J. Expert Syst. Appl. **38**(5), 5311–5319 (2011)
6. Ozaki, T.: On the order determination of ARIMA models. J. R. Stat. Soc. Ser. C **26**(3), 290–301 (1977)
7. Platt, J.C.: Sequential minimal optimization: A fast algorithm for training support vector machines. Technical report MSR-TR-98-14, Microsoft Research (1995)
8. Yu, L., Wang, S., Lai, K.K.: Mining stock market tendency using GA-based support vector machines. In: International Workshop on Internet and Network Economics (WINE). LNCS, vol. 3828, pp. 336–345 (2005)

Do Information Quantity and Transmission Make a Difference to the Stable Contrarian?

Hung-Wen Lin[1], Jing-Bo Huang[2], Kun-Ben Lin[1],
and Shu-Heng Chen[3]([envelope])

[1] Nanfang College of Sun Yat-sen University, Guangzhou, China
[2] Sun Yat-sen University, Guangzhou, China
[3] National Chengchi University, Taipei, Taiwan
chen.shuheng@gmail.com

Abstract. We study how financial transparency and media coverage work in the Chinese stock markets. In this paper, transparency means information quantity, while media means information transmission. The market has negative momentum profits no matter how transparency or media coverage changes, which suggests that transparency or media coverage does not work individually in China. High transparency and high media coverage make significantly positive momentum profits, whereas low transparency and low media coverage make significantly negative momentum profits. These outcomes show that transparency and media coverage work jointly in China. Our findings imply that information quantity and transmission are both crucial in China.

Keywords: Transparency · Media coverage · Momentum

1 Introduction

As is known to all, China is the second largest economy in the world. It increasingly has an impact on the world economy. Since the establishments of Shanghai stock exchange and Shenzhen stock exchange in 1990s, the Chinese stock markets have been absorbing the capital from all over the world. Consequently, it is necessary to study these markets. We focus on the financial transparency and media coverage in the Chinese stock markets.

Financial transparency is a hot topic in the accounting research. The United Nations conference on trade and development (UNCTAD) has defined the transparency as a procedure making the corporate status, decisions and activities available. We use transparency to stand for information quantity in this paper. Our involvement of transparency also comes from its relationships with stock prices in the literature. For example, the less transparent stocks are easy to face price crashes (See Hutton et al. 13). Firth et al. (9) have proved that transparency can reduce the effects of investor sentiment on stock prices. Furthermore, transparency may affect the market states. Johannesen and Larsen (15) perceived that transparency with respect to sensitive information may lead to stock market loss. Hence, our proposal of transparency is reasonable.

There are a lot of retail investors in China. Barber and Odean (2) have provided evidence that individual investors mainly rely on media to acquire news and they are

© Springer Nature Switzerland AG 2019
E. Bucciarelli et al. (Eds.): DCAI 2018, AISC 805, pp. 9–17, 2019.
https://doi.org/10.1007/978-3-319-99698-1_2

likely to trade the stocks in news. Thus, it is indispensable to study media coverage in China. Intuitively, media is an important channel to disseminate information, so we employ media coverage as the proxy of information transmission. In the literature, the connections of media with stock prices and returns have been documented. Tetlock et al. (18) held that the tones in news from media can predict the future stock returns. In addition, stock prices react strongly to the news from media (See Dyck and Zingales 8). Based on the above discussions, we confidently recommend media coverage due to its existing dissections.

Jegadeesh and Titman (14) have found significant momentum in the US stock market. They believe that momentum stems from the delayed reactions to firm specific information. From the standpoint of information, momentum has also been studied, such as Chan et al. (4), Johnson (16) and so on. We deem that transparency and media coverage may also be related to momentum when the information is the medium among them.

To make the working patterns of transparency and media clear, we construct a number of momentum portfolios by the independent sorts and sorts with specific orders. As for two-way sorted momentum portfolios, we generate them by sorts on transparency (media coverage) and returns. These portfolios aim at the individual patterns of transparency or media coverage. For the joint pattern of transparency and media coverage, we create three-way sorted momentum portfolios from the sorts on the above three variables.

We organize the remainder as follows. The descriptions of our model and data are in Sect. 2. In Sect. 3, the empirical outcomes of two-way sorted momentum portfolios and three-way sorted momentum portfolios appear. We check the robustness in Sect. 4. Finally, Sect. 5 concludes this paper.

2 Models and Data

According to Bhattacharya et al. (3), we calculate earnings aggressiveness, earnings smoothing and loss avoidance for transparency. The measure for media coverage derives from Hillert et al. (12). Jegadeesh and Titman (14)'s process on momentum profit is exerted.

2.1 Transparency and Media Coverage Computations

We use the decile rankings on earnings aggressiveness, earnings smoothing and loss avoidance of the stocks. The average decile of a stock is its transparency. Discretionary accruals represent earnings aggressiveness, which are the difference between total accruals scaled by total assets and nondiscretionary accruals (Dechow et al. 6). McInnis (17)'s methodology is used to compute earnings smoothing. We ponder that loss avoidance and timeliness of loss recognition are negatively correlated. Logically, when the earnings losses are recognized more timely, the losses have lower possibility to be avoided. The procedure of calculating loss recognition follows Balla and Shivakumar (1). Some adjustments regarding the Chinese market are exerted to the media model of

Hillert et al. (12). We use the CSI 300 and Shenzhen dummy to take the places of S&P 500 and NASDAQ dummy of Hillert et al. (12).

2.2 Data Description

The financial data and media data are collected from the China Stock Markets and Accounting Research (CSMAR) database and China Infobank database respectively. The period of our data begins from 2005 to 2016 with seasonal basis, including all the stocks in Shanghai stock exchange and Shenzhen stock exchange.

The Chinese market significantly has a total average momentum profit of −0.026 (*t*-stat is −3.309) with 2-season formation and holding periods. We can see that the Chinese stock market is contrarian instead of momentum. The total average media coverage is insignificantly positive of 0.004, and transparency is also insignificantly positive of 5.479 (*t*-stats are 0.103 and −0.381 respectively). These results suggest that transparency and media coverage are in normal ranges during our sample period.

3 Empirical Outcomes

We produce two-way sorted momentum portfolios and three-way sorted momentum portfolios in this section. The independent sorts and the sorts with specific orders are both used for the generations of momentum portfolios.

3.1 Two-Way Sorted Momentum Portfolios

In order to study how transparency or media coverage works individually in the market, we construct 6 two-way sorted momentum portfolios and calculate their momentum profits. If the momentum profits stay negative in spite of the individual changes of transparency or media coverage, it means that transparency or media coverage does not work individually (Table 1).

All the momentum profits from the portfolios sorted by transparency and stock returns are negative. For instance, *Spec (R, T)* has significantly negative momentum profits with high transparency, second level transparency (*t*-stats are −2.093 and −2.204 respectively). Within *Spec (T, R)*, the significantly negative momentum profits come with high transparency and low transparency (*t*-stats are −2.190 and −1.694 respectively). These outcomes show that the market still has significantly negative momentum profits regardless of the most of changes in transparency, so transparency does not work individually in the Chinese contrarian market.

Although the most of momentum profits from the portfolios sorted by media coverage and stock returns are insignificant, all the momentum profits are negative. *Spec (C, R)* has a significantly negative momentum profit with low media coverage (*t*-stat is −1.914). We can observe that the market still has negative momentum profits no matter how the media coverage changes. In other words, media coverage does not work individually in the Chinese contrarian market.

Table 1. Two-way sorted momentum portfolios

Panel A: Indep (R, T)		Panel B: Indep (R, C)		Panel C: Spec (R, T)	
T	MP	C	MP	T	MP
H	−0.019**	H	−0.005	H	−0.018**
2	−0.01	2	−0.012	2	−0.016**
3	−0.011	3	−0.014	3	−0.005
L	−0.012	L	−0.016	L	−0.013
Panel D: Spec (R, C)		Panel E: Spec (T, R)		Panel F: Spec (C, R)	
C	MP	T	MP	C	MP
H	−0.011	H	−0.018**	H	−0.018
2	−0.01	2	−0.008	2	−0.02
3	−0.013	3	−0.012	3	−0.013
L	−0.019	L	−0.013*	L	−0.025*

Presented are the momentum profits (MP) from two-way sorted momentum portfolios. The formation and holding period are 2 seasons. We use some notations for the portfolios. For example, Indep (X, Y) is from the independent sorts on X and Y. Spec (X, Y) is from the first sort on X and second sort on Y. The stock return is R, transparency is T, and media coverage is C. The sorts on these three variables for the stocks are 4 groups with descending orders. The high level is H and low level is L. *, **, *** represent significance at 10%, 5% and 1% respectively.

3.2   Three-Way Sorted Momentum Portfolios

For the sake of understanding how transparency and media coverage work jointly in the market, we create 5 three-way sorted momentum portfolios and compute their momentum profits. When the momentum profits remain negative regardless of the joint changes of transparency and media coverage, they do not work jointly.

From Table 2, we can draw from the above outcomes that high transparency and high media coverage make momentum, whereas low transparency and low media coverage make contrarian. For example, Spec (R, T, C) significantly has a positive momentum profit with high transparency and high media coverage (t-stat is 1.943), while it has a significantly negative momentum profit with low transparency and low media coverage (t-stat is −1.679). Moreover, Spec (R, C, T) has same results. These outcomes show that transparency and media coverage can work jointly in Chinese contrarian stock market.

High transparency and high media coverage create a condition of much information and wide information transmission. The investors will not have enough time to process information, so they are easy to believe the information. Some researchers think the Chinese investors have obvious herding behaviors (See Gong and Dai 10; Ulzii et al. 19). They intend to form same opinions. These opinions are also possibly preserved under high transparency and high media coverage. They buy winner stocks and sell loser stocks. The prices of winner stocks keep increasing and the prices of loser stocks keep decreasing. Therefore, momentum takes place.

Table 2. Three-way sorted momentum portfolios

Panel A: *Indep* (R, T, C)		Panel B: *Spec* (R, T, C)		Panel C: *Spec* (R, C, T)	
C,T	*MP*	*T, C*	*MP*	*C, T*	*MP*
H, H	0.007	H, H	0.037*	H, H	0.042***
H, M	0.018	H, M	−0.013	H, M	−0.008
H, L	−0.004	H, L	0.027	H, L	0.007
M, H	−0.003	M, H	0.002	M, H	0.013
M, M	−0.017	M, M	0.011	M, M	−0.013
M, L	0.007	M, L	−0.005	M, L	−0.013
L, H	0	L, H	−0.005	L, H	0.008
L, M	−0.003	L, M	−0.005	L, M	0.003
L, L	−0.034**	L, L	−0.022*	L, L	−0.029**

Panel D: *Spec* (T, C, R)		Panel E: *Spec* (C, T, R)	
T, C	*MP*	*C, T*	*MP*
H, H	0.023	H, H	0.022
H, M	0.012	H, M	0.025
H, L	−0.008	H, L	0.001
M, H	0.019	M, H	0.009
M, M	−0.02	M, M	−0.001
M, L	0.003	M, L	0.005
L, H	−0.013	L, H	−0.005
L, M	0.005	L, M	−0.001
L, L	−0.018	L, L	−0.025

Presented are the momentum profits (*MP*) from three-way sorted momentum portfolios. The formation and holding period are 2 seasons. We use some notations for the portfolios. For example, *Indep (X, Y, Z)* is from the independent sorts on *X, Y,* and *Z. Spec (X, Y, Z)* is from the first sort on *X*, second sort on *Y*, and third sort on *Z*. Due to the sample size, the sorts on these three variables for the stocks are 3 groups with descending orders. The high level is *H*, middle level is *M* and low level is *L*. *, **, *** represent significance at 10%, 5% and 1% respectively.

Low transparency and low media coverage bring little information and narrow information transmission. DeBondt and Thaler (7) have found that the investors will overreact to unexpected news. This circumstance lets the investors stay in the murk. When the present information is out of the investors' expectations, the investors adjust the past decisions according to the present information. They will also overreact to the information. Consequently, contrarian occurs.

4   Robustness Checks

The *Spec (T, R, C)* and *Spec (C, R, T)* are not studied in Sect. 3. We give the results in this section. We also inspect *Indep (R, T, C)*, *Spec (R, T, C)*, *Spec (R, C, T)*, *Spec (T, C, R)*, and *Spec (C, T, R)* in the extensive periods. The formation and holding period are extended from 2 to 12 seasons.

4.1   *Spec (T, R, C)* and *Spec (C, R, T)*

In this section, *Spec (T, R, C)* and *Spec (C, R, T)* are studied with 2-season formation and holding period. We find that the joint working pattern of transparency and media coverage in 2-season period is robust. For instance, *Spec (T, R, C)* has a significantly negative momentum profit with low transparency and low media coverage (*t*-stat is −5.724). By *Spec (C, R, T)*, a significantly positive momentum profit arises with high transparency and high media coverage (*t*-stat is 2.692).

4.2   Extensive Periods

The formation and holding period of *Indep (R, T, C)*, *Spec (R, T, C)*, *Spec (R, C, T)*, *Spec (T, C, R)* and *Spec (C, T, R)* are prolonged from 2 seasons to 12 seasons. The Figs. 1 and 2 show the outcomes of *Indep (R, T, C)* and *Seq (R, T, C)*. The outcomes of *Spec (R, C, T)*, *Spec (T, C, R)* and *Spec (C, T, R)* are very similar with the followings, so we do not show their outcomes because of the length of this paper.

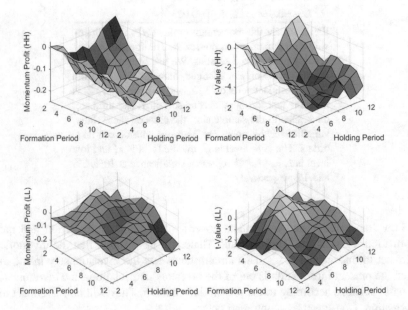

Fig. 1. Outcomes of *Indep (R, T, C)* with extensive formation and holding period.

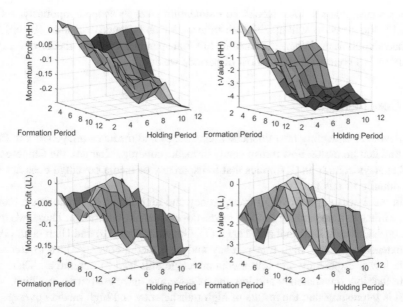

Fig. 2. Outcomes of *Seq (R, T, C)* with extensive formation and holding period.

From Figure 1, we can see that the most of momentum profits turn to be negative in the extensive periods. There are some positive momentum profits, but a lot of them are insignificant. For instance, *Indep (R, T, C)* has an insignificantly positive momentum profit of 0.015 (*t*-stat is 0.378) with high transparency and high media coverage in the 2-season formation and 11-season holding period. By low transparency and low media coverage, it has an insignificantly positive momentum profit of 0.011 (*t*-stat is 0.211) in the 3-season formation and 6-season holding period.

In the extensive periods, as can be seen from Figure 2, the most of momentum profits are negative though some insignificantly positive momentum profits exist. With 4-season formation and 3-season holding period, it has an insignificantly positive momentum profit of 0.023 (*t*-stat is 0.712) by high transparency and high media coverage. It also has an insignificantly positive momentum profit of 0.001 (*t*-stat is 0.046) with 5-season formation and 5-season holding period via low transparency and low media coverage.

4.3   Implications for the Extensive Outcomes

Under high transparency and high media coverage, it is evident that momentum in 2-season formation and holding period easily turns to contrarian in the extensive periods. Compared to those in America and Europe, the financial statements of Chinese corporations are relatively less transparent (e.g. Habib and Jiang 11). Therefore, the information of Chinese corporations is possibly unreliable though the information is sufficient and widely transmitted. However, the Chinese investors are lack of enough knowledge and experience (e.g. Chen et al. 5). In the 2-season period, the investors do not process the information properly, but have same opinions (e.g. Gong and Dai 10; 19). They buy winner stocks and sell loser stocks, so momentum takes its shape temporarily. As time goes by, the investors will update their information and change decisions. Accordingly, it makes contrarian. When it comes to low transparency and low media coverage, it stably makes contrarian in the most of periods we adopt.

5 Conclusions

We devote to dissecting how financial transparency and media coverage work in China. We find that no matter how transparency or media coverage changes, the Chinese stock market stays contrarian. It implies that transparency or media coverage does not work individually in this market.

In the 2-season period, high transparency and high media coverage make momentum, while low transparency and low media coverage make contrarian. These two results are consistent with DeBondt and Thaler (7), Gong and Dai (10) and Ulzii et al. (19). In the extensive periods, high transparency and high media coverage turn to contrarian, while low transparency and low media coverage still lead to contrarian. There is no doubt that low transparency and low media coverage make contrarian in China.

It is interesting that the results of high transparency and high media coverage vary along with time. Generally speaking, the financial statements of Chinese corporations are not very transparent and the Chinese investors do not have abundant knowledge and skills (e.g. Habib and Jiang 11; 5). The ample and widely transmitted information may be unreliable but trusted by the investors. The investors are very likely to form same opinions and decisions, so they buy winner stocks and sell loser stocks. Due to this consequence, the prices of winners keep ascending and the prices of losers keep descending. Finally, momentum appears temporarily during 2-season period. Gradually, the investors update their information as time goes by, which may urge them to change decisions. Thus, contrarian takes the place of momentum.

The Chinese stock market is quite different from the mature stock markets in America and Europe. It has extremely significant contrarian instead of momentum, suggesting that the prices of stocks are easy to reverse. If the investors form their decisions based on the perspectives of price inertia, the investments will meet evident risks. We deem that the insider operation of corporations may be one of the reasons for unreliable information and relatively less transparent financial statements in China. This kind of information will also lead to obvious uncertainties and risks. All the possibilities we mentioned above imply that momentum may relate to risk studies. The researches with the involvements of price risks and insider operations will reveal some new insights for this field.

References

1. Balla, R., Shivakumar, L.: Earnings quality in UK private firms: comparative loss recognition timeliness. J. Account. Econ. **39**, 83–128 (2005)
2. Barber, B.M., Odean, T.: All that glitters: the effect of attention and news on the buying behavior of individual and institutional investors. Rev. Financ. Stud. **2**, 785–818 (2008)

3. Bhattacharya, U., Daouk, H., Welker, M.: The world price of earnings opacity. Account. Rev. **78**, 641–678 (2003)
4. Chan, L.K.C., Jegadeesh, N., Lakonishok, J.: Momentum strategies. J. Financ. **51**, 1681–1713 (1996)
5. Chen, G., Firth, M., Gao, N.: The information content of concurrently announced earnings, cash dividends, and stock dividends: an investigation of the chinese stock market. J. Int. Financ. Manag. Account. **13**, 101–124 (2002)
6. Dechow, P.M., Sloan, R.G., Sweeney, A.P.: Detecting earnings management. Account. Rev. **70**, 193–225 (1995)
7. DeBondt, W., Thaler, R.: Does the stock market overreact? J. Financ. **3**, 793–805 (1985)
8. Dyck, A., Zingales, L.: The Media and Asset Prices, Working Paper, Harvard Business School (2003)
9. Firth, M., Wang, K., Wong, S.M.: Corporate transparency and the impact of investor sentiment on stock prices. Manag. Sci. **61**, 1630–1647 (2015)
10. Gong, P., Dai, J.: Monetary policy, exchange rate fluctuation, and herding behavior in the stock market. J. Bus. Res. **76**, 34–43 (2017)
11. Habib, A., Jiang, H.: Corporate governance and financial reporting quality in China: a survey of recent evidence. J. Int. Account. Audit. Tax. **24**, 29–45 (2015)
12. Hillert, A., Jacobs, H., Müller, S.: Media makes momentum. Rev. Financ. Stud. **27**, 3467–3501 (2014)
13. Hutton, A.P., Marcus, A.J., Tehranian, H.: Opaque financial reports, R2, and crash risk. J. Financ. Econ. **94**, 67–86 (2009)
14. Jegadeesh, N., Titman, S.: Returns to buying winners and selling losers: implications for stock market efficiency. J. Financ. **48**, 65–91 (1993)
15. Johannesen, N., Larsen, D.T.: The power of financial transparency: an event study of country-by-country reporting standards. Econ. Lett. **145**, 120–122 (2016)
16. Johnson, T.: Rational momentum effects. J. Financ. **57**, 585–608 (2002)
17. McInnis, J.: Earnings smoothness, average returns, and implied cost of equity capital. Account. Rev. **85**, 315–341 (2010)
18. Tetlock, P.C., Saar-Tsechansky, M., Macskassy, S.: More than words: quantifying language to measure firms' fundamentals. J. Financ. **3**, 1437–1467 (2008)
19. Ulzii, M., Moslehpour, M., Kien, P.V.: Empirical models of herding behaviour for asian countries with confucian culture. In: Predictive Econometrics and Big Data. Studies in Computational Intelligence, vol. 753 (2018), https://doi.org/10.1007/978-3-319-70942-0_34

The Logistic Map: An AI Tool for Economists Investigating Complexity and Suggesting Policy Decisions

Carmen Pagliari[✉] and Nicola Mattoscio

Department of Philosophical, Pedagogical and Economic-Quantitative Sciences,
University of Chieti-Pescara, Viale Pindaro 42, 65127 Pescara, Italy
{carmen.pagliari,nicola.mattoscio}@unich.it

Abstract. The present contribution contains an original interpretation of the logistic map popularized by the biologist Robert May in 1976. This map is potentially a powerful AI tool based on a deterministic methodology having a double possibility to be applied in economics. The first application is to investigate the intrinsic complexity of real economic phenomena characterized by endogenous non-linear dynamics. The second application is to determine results, typical of a normative science, useful for suggesting policy decisions aimed to avoid chaos and unpredictability in the real economic system. In the first type of application, the logistic map can be used as an AI tool of forecasting (for previsions of bifurcations, cycles and chaos). In the second, the logistic map can be considered as an AI tool for policy makers in order to deduce the analytical conditions that ensure the economic system to be sufficiently far away from chaos and uncontrollability.

Keywords: Economic non-linear dynamics · Complexity · Chaos
Policy decisions · Logistic map

1 Introduction

"Catastrophe theory" emerged and developed in the 1960s and 1970s, having initially been presented to the scientific community in 1972 by the French mathematician René Thom in his book, *Structural Stability and Morphogenesis* [27][1]. Thom's results achieved success among the scholars studying the structural dynamics of real systems in various scientific fields. Specific features of the approach included analyses of bifurcations and periodic cycles and the study of deterministic chaos in nonlinear contexts.

In the 1980s, numerous works showed the importance of nonlinear dynamic analysis in interpreting economic phenomena, particularly economic cycles. (See, among others, Baumol and Benhabib [4], Day [10], Farmer [13], Grandmont [15] and Medio [24]). Part of this literature worked on identifying causes of irregularities in

---

[1] The term "catastrophe" refers to a sudden change in a system's developmental trajectory, or in a variable that characterizes it, caused by small changes in the initial conditions and/or in conditions outside the system.

© Springer Nature Switzerland AG 2019
E. Bucciarelli et al. (Eds.): DCAI 2018, AISC 805, pp. 18–27, 2019.
https://doi.org/10.1007/978-3-319-99698-1_3

fluctuations in the macroeconomic variables involved. (See, among others, Barkley Rosser [3], Benhabib [5], Benhabib and Day [6], Day [8, 9], Grandmont [16, 17] and Nishimura and Sorger [25]).

Understanding what causes these fluctuations is crucial if we are to control such irregularities, which lead to two very serious problems for researchers and policy makers: unpredictability in how the systems or economic phenomena develop over time, and their uncontrollability.

The unpredictability of an economic phenomenon's evolution is not the result of random events or random factors only. In many cases, it is the result of dynamic processes present intrinsically within the system, which are unknown or poorly understood because of the inadequacy of current methods of inquiry.

Our perspective is based on the conviction that, generally, the dynamic trajectory of a real phenomenon can be broken down into several components, some deterministic and some stochastic. The first of these can be studied precisely, as long as we have the appropriate scientific and/or technological means. The second can be investigated by statistical and probabilistic tools. Both types of component deserve attention from researchers in the applied sciences, even if understanding them requires profoundly different methodologies. Within economics, the most common error regarding these two approaches to economic dynamics is to view the deterministic approach as less useful for understanding the phenomena observed than the stochastic approach. The arguments of Faggini and Parziale are interesting in this regard [12].

Our pronouncement in favor of the first of these approaches is driven by the belief that understanding the temporal evolutions of the deterministic components of an economic phenomenon lends support to the research of stylized facts characterized by coherence with reality.

In particular, in an initial step of investigation, such an understanding allows us to measure the degree of complexity within the observed phenomenon. In a subsequent step, by the analysis of the deterministic dynamics, we can investigate the circumstances that have to be avoided in order to escape from the uncontrollability of the fluctuations present in the observed phenomenon.

In this paper, we propose an original interpretation of the logistic map, the best known and simplest tool for deterministic nonlinear dynamic analysis, popularized by the biologist Robert May in 1976 [23][2]. We aim to highlight the logistic map's possible role as an analytical tool for studying real economic phenomena, evaluating their degree of complexity, and determining the analytical conditions for convergence and stability.

We consider the logistic map under the umbrella of AI for the following reasons. Firstly, it is a one-dimensional recursive map, and so it has memory of the past over time, since each value of the variable is calculated on the basis of the previous one; for this reason, it is a learning function. Moreover, despite its simplicity, it is rich of

---

[2] The logistic map is a unimodal polynomial map of degree 2, defined for the closed and limited interval of real numbers [0,1]. It is typically used to study the onset of bifurcations, cycles, and deterministic chaos in situations characterized by simple nonlinear dynamics and by limits on the variable whose temporal evolution is to be analyzed.

information about the future evolution of the variable whose dynamics depends on the characteristic parameter of the map and on the initial conditions [21].

In particular, the logistic map can be considered as an AI tool for artificial investigations on the dynamics of real systems in different scientific areas (See, among others, [28]). For example, the logistic map is used for artificial simulations of perturbations in the trajectory of a variable by means of changes in its characteristic parameter; in such a way, it can be a tool for simulating regular or chaotic changes in models of applied sciences [14].

The logistic map is also used as an algorithm for obtaining artificial time series [11], and for generating chaotic variables with the nature of pseudo-random numbers [26]. In relation to the role of pseudo-random numbers generator, the logistic map can be considered at the base of chaos search [21].

Our main aim is to offer policy makers a simple way of identifying the ranges of variability, for opportune variables, within which chaos and uncontrollability of the system—in contexts characterized by the absence of random events—can be avoided. Interesting studies relevant to this approach include those by Bullard and Butler [7], Holyst [18], Holyst and Urbanowicz [19], Kaas [20], and Kopel [22]. Section 2 presents a specific analytical formulation of the logistic map that can be applied to economic contexts. Section 3 gives a summary of the analytical treatment of the conditions for compatibility and stability of the alternative temporal paths of the variables in question. The main implications for economic policy deriving from these conditions are given in Sect. 4.

2   An Analytical Setting of the Logistic Map for Economic Contexts

Our aim here is not to present the traditional analytical treatment of the logistic map (see [1], Chap. 1); instead, we offer a formal setting of this map that lends itself to the study of complex economic phenomena.

Consider two economic variables y and x that characterize a real phenomenon whose evolution over time we want to analyze. Both are dependent on the time variable t, which is treated as a discrete variable. Let y_t be correlated linearly to x_t. That is

$$y_t = \delta_C x_t \quad \forall t \quad (\delta_C > 0 \text{ or } \delta_C < 0). \tag{1}$$

A relation of the type (1) is very common and it can be interpreted as a linear relation between the values that the variables take on over the same period. For example, such relations can be derived from the definitions of the two variables, or from historically consolidated statistical correlations insofar as they correspond to the rules of economic policy, or to the behavior of the economic agents.

Relation (1) is useful because it allows one to focus attention on the temporal evolution of variable x alone, and then immediately deduce that of the related variable y.

Suppose that the two variables are also causally related in a dynamic context that contemplates at least two periods and a delay of one period.

To see this, let y be an effect of x. In particular, we consider that the value of y at time $(t + 1)$ is the result of the algebraic sum of two effects on y caused by the value of x at time t. These two effects are called direct effect $D(x_t)$ and indirect effect $I(x_t)$:

$$y_{t+1} = D(x_t) + I(x_t) \quad \forall t. \tag{2}$$

Suppose that the direct effect is linear with respect to the variable x, while the indirect effect is quadratic with respect to the same variable:

$$D(x_t) = \delta_D x_t \quad \forall t \quad (\delta_D > 0 \text{ or } \delta_D < 0) \tag{3}$$

$$I(x_t) = \delta_I x_t^2 \quad \forall t. \tag{4}$$

The linearity of the direct effect is plausible if we take (1) into account and consider $D(x_t)$ as the result of a linear regression. The nonlinearity of the indirect effect is justified because it contributes to the value of y deriving from the direct linear effect of a third variable that, in turn, depends linearly on the same value x_t.

Note that δ_C is the coefficient of the current linear relationship between the two variables, while δ_D is the coefficient of the delayed relationship.

An interesting example of a quadratic relationship between economic variables comes from macroeconomic analyses of the interdependence between the output gap and the unemployment rate. The quadratic dependence can be explained by observing that a variation in the unemployment rate leads directly to a variation in aggregate income, and then influences this last variation indirectly (with a period's delay) through a variation in aggregate consumption spending, that is a component of aggregate income. In other words, any change in the unemployment rate also has an indirect effect on GDP by means of the variation in a component included in the direct effect on GDP and, therefore, it has a quadratic impact on the change in the GDP growth rate and on the output gap.

A second example of quadratic dependence comes from examining the relationship between the deficit/GDP ratio and the debt/GDP ratio under austerity. The indirect effect on the deficit/GDP ratio of an increase in the debt/GDP ratio, itself correlated with the former, comes from a subsequent increase in debt/GDP ratio, caused by a fall in GDP due to austerity policies that arise from the debt/GDP ratio's own tendency to grow. The debt/GDP ratio therefore has a quadratic effect on the deficit/GDP ratio.

In analytical formulations of these two examples, the coefficients δ_C and δ_D may depend on variables with particular economic significance.

In the first example, these coefficients can be related, respectively, to a labor force productivity index (non-delayed effect) and to a labor market flexibility index (delayed effect).

In the second example, the coefficient of the non-delayed relationship can be considered as unitary; the coefficient of the delayed relation can be seen as negatively correlated with the GDP growth rate, and positively correlated with the average nominal interest rate on government bonds.

In both examples and their applications, policy makers should aim to maintain control over these coefficients.

Taking the relations (1)–(4) simultaneously, with the goal of producing an equation containing only the variable x, we obtain:

$$x_{t+1} = (\delta_D/\delta_C)x_t + (\delta_I/\delta_C)x_t^2. \tag{5}$$

By performing the substitutions $(\delta_D/\delta_C) = \lambda$ and $(\delta_I/\delta_C) = \Lambda$, we obtain:

$$x_{t+1} = \lambda x_t + \Lambda x_t^2. \tag{6}$$

Applying the following analytical transformation:

$$x_t = -(\lambda/\Lambda)X_t \tag{7}$$

we obtain the canonical form of the logistic map with the normalized auxiliary variable X_t (the two boundary cases are negligible and, therefore, the analysis is carried out for $0 < X_t < 1$, $\forall t$):

$$X_{t+1} = \lambda(1 - X_t)X_t \quad \forall t. \tag{8}$$

The variable under consideration is x_t. Transformation (7) easily allows us to trace its temporal path back to that of the auxiliary variable X_t.

We see that $(\delta_D/\delta_C) = \lambda$ must be a positive number, having assumed that the non-zero coefficients δ_D and δ_C have the same sign.

Because X_t is a normalized variable, the characteristic parameter λ must also be less than 4. This can be deduced from the phase diagram $X_{t+1} = f(X_t)$ of map (8), where the ordinate of the vertex of the generic arc of parabola is equal to $\lambda/4$, which must be less than 1.

Note, finally, that the characteristic parameter λ of the logistic map in its canonical form, based on the formulation we have proposed, is equal to the ratio between the coefficient of the linear effect of x's current value on y with the delay of one period and the coefficient of the undelayed linear relation between the current values of y and x.

Holyst and Urbanowicz [19] offer an interesting contribution to understanding the dynamics in cases where there is a one-period delay, and the related implications for economic policy.

3   The Logistic Map's Analytical Properties: Stability, Cycles, and Intrinsic Disorder (Chaos)

In this section, we present some of the logistic map's analytical properties that are useful for our purposes here.

The analytical treatment of the stability of the solutions to Eq. (8), when the two boundary cases $X_t = 0$ and $X_t = 1$ are excluded and the characteristic parameter lies within the analytically compatible range of real numbers $(0,4)$, synthetically implies the following results (see [1], Chap. 1):

- for $0 < \lambda \leq 1$, the temporal path of the variable X_t converges toward a zero stationary state;
- for $1 < \lambda < 3$, the temporal path of the variable X_t converges toward a non-zero stationary state;
- for $3 \leq \lambda < 3.4495...$, the temporal path of the variable X_t undergoes a first bifurcation (two-period cycle);
- for $3.4495... \leq \lambda < 3.5699...$, the bifurcations of the temporal path of the variable X_t progressively double as λ grows (cycles of increasing frequency);
- for $3.5699... \leq \lambda < 4$, we observe the presence of cycles characterized by "infinite periodicity" in the temporal path of the variable X_t; this is equivalent to the absolute unrepeatability of any value assumed by the variable itself, whose evolution appears stochastic, but is only absolutely disordered, *i.e.* deterministic chaos arises.

As we saw in Sect. 2 above, the parameter λ is equal to the ratio between the coefficient of the linear effect of the current value of x on y with a one-period delay and the coefficient of the undelayed linear relation between the current values of y and x. From this analytical formulation of the characteristic parameter and from the results presented above, useful conclusions can be deduced.

In fact, the convergence of the temporal path of the variable studied to a dynamically stable solution determines whether $1 < \lambda < 3$, that is equivalent to $\delta_C < \delta_D < 3\delta_C$.

This is equivalent to saying that the stability of the temporal path of the variable to be controlled is guaranteed, as long as the coefficient of the delayed linear effect is greater than the coefficient of the non-delayed relation between y and x and is no higher than three times it. In fact, this overrun leads to bifurcations and cycles of increasing periodicity, until the onset of intrinsic disorder—that is, until the onset of chaos.

The stability of the temporal evolution of x is possible in the presence of the combinations (δ_C, δ_D) that belong to the convergence zone of Fig. 1.

To avoid chaos, $(3.5699...)\delta_C < \delta_D < 4\delta_C$ must not be the result—that is, the pair (δ_C, δ_D) must not correspond to a point within the chaos zone of Fig. 1.

We observe that the combinations of the coefficients δ_C and δ_D that can ensure the stability of the temporal path of the variable under consideration are infinite, but the intervals of simultaneous variability that ensure convergence toward a stable non-zero stationary state are well determined.

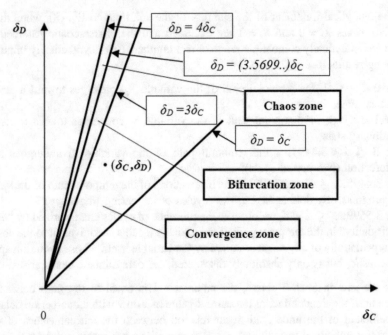

Fig. 1. Graphical representation of the AI tool

4 An AI Tool for Evaluating and Predicting Complex Economic Phenomena and Economic Policy Decisions

Based on the analytical formulation proposed above, the possibility of identifying the zone to which the pair (δ_C, δ_D) belongs (Fig. 1) is doubly useful. First, it provides a criterion for immediately verifying the stability of the dynamics studied and a tool for predicting the onset of bifurcations, cycles, and intrinsic disorder (chaos) when the combinations of the two coefficients δ_C and δ_D vary. Secondly, it allows us to easily calculate how close we are to the uncontrollability arising from deterministic causes intrinsic to the system, and to determine the range of variability for the combinations of coefficients within which it is possible to avoid chaos. In our opinion, this tool's dual role may be very important for policy decisions. We believe it could be considered as an artificial intelligence tool for economists who want to study the complexity of economic phenomena and their implications for economic policy decisions.

The examples of possible economic applications of our dynamic nonlinear model given in Sect. 2 allow several remarks in support of this. Our first example looked at the dynamic analysis of the effects of changes in the unemployment rate on the output gap. For economic policy decisions, the combination of the two coefficients (δ_C, δ_D)—labor force productivity (non-delayed effect) and labor market flexibility (delayed effect)—are crucially important. Economic policy decisions aimed to avoid the instability and uncontrollability of occupational phenomena arising from causes intrinsic to the system must ensure this pair of coefficients lies within the convergence zone of our

AI tool's graphical representation (Fig. 1). Note that there is an infinite number of combinations of the two control coefficients (δ_C, δ_D) that can ensure the stability and predictability of the dynamics studied by policy makers. This analytical aspect provides significant margins of choice and tolerance for economic policy decisions. Furthermore, we can easily estimate the distance of the current combination (δ_C, δ_D) from the chaos zone—that is, it is possible to predict what margin of tolerance exists for current or planned economic policy, and so avoid the chaos zone and the uncontrollability and unpredictability it inevitably brings. A similar interpretation is available for the second example in Sect. 2. Applying our AI tool in the same way, the first element of the pair (δ_C, δ_D) is unitary; the second is negatively correlated with the GDP growth rate and positively with the average nominal interest rate on government bonds. The analytical definition of these coefficients implies that useful conclusions for economic policy can be expressed by combinations of GDP growth rate and interest rates on government bonds that will guarantee the stability of the temporal evolution of the debt-GDP ratio.

5 Conclusions

This paper presents an easy-to-use tool for exploring the complexity of the economic phenomena whose determining variables policy makers aim to control. The goal of our proposal is closely related to the analysis of the dynamics of the deterministic components of the system's temporal evolution. Typically these components are neglected in the studies of real economic phenomena, because it is widely believed that their effects are dominated by stochastic components, which are often present in large numbers. In our opinion, this does not justify the almost total abandonment of efforts to discover the characteristics of the inscrutable, ineluctable forces within the system that silently follow deterministic laws. It is as though researchers into earthquakes were to study seismic phenomena only from a statistical and probabilistic point of view, abandoning any inquiry into the Earth's geological dynamics because they lacked the investigative tools.

We should perhaps think more closely about the fact that the deterministic component of a real system's dynamic evolution—particularly that of an economic system—needs to be understood if we are to *effectively* pursue its positive characteristics, viewing them as so many features to be preserved, and simultaneously to *really* neutralize its negative characteristics by compensating for them in every way possible, treating them as so many harmful features.

Suppose that, on a purely biological basis, the number of heartbeats available to each individual could be predetermined by looking at the biochemical formula of their DNA. We would hardly be indifferent to this fact. And it would be equally disturbing to think that a reduction in the effective number of heartbeats was the result of multiple causes, dependent on individual lifestyle, events related to the individual's health, traumatic events, and other external relational and environmental factors. All of us would want to know how big our predetermined allocation of heartbeats was, in order to use all of it, without wasting a single one. This should be the attitude of the economists researching the intrinsic laws of the dynamics of an economic system.

The limitations of our study refer mainly to the choice of a pure deterministic approach and to the analytical simplicity of the map considered. Both limitations, although restrictive for the applications to real economic phenomena, are, in our opinion, very useful to the progress of AI tools for economic studies. The first limitation allows us to highlight the role of the study of deterministic laws hidden in the economic systems. The second limitation is apparent, because the simplicity of the logistic map enlarges the potentialities of this tool in relation to chaos search and rends this learning function an easy-to-use instrument for simulating and studying complex phenomena. These last two implications—deriving from the simplicity of the map considered—are the main features that can support the qualification of the logistic map as an AI tool. Furthermore, we suggest to reflect on the prospect that Chaos Theory can provide a bridge between deterministic and statistic behaviour. In particular, this result can be obtained using the logistic map as a generator of aperiodic series of random numbers, a valuable tool for computer simulation methods (See, among others, [2, 21, 26]). The importance of chaos search for the study of real phenomena in a complex system, is perceived by scholars of different applied sciences and, therefore, we expect that chaos search—and hence the logistic map—will become more present in AI applications.

Finally, we try to provide suggestions aimed at determining if the predictability of the logistic map remains economically as well as statistically significant when applying it to real data. Our opinion is that the predictive ability of logistic map, if applied to real economic data, may provide information about the trend of the variable in dependence on the intrinsic laws of the system, if random events are negligible. When such events are not negligible, this predictability—although not anymore meaningful for the prevision of the temporal evolution of the economic variable considered—remains a significant support to the identification of economic stylized facts in complex dynamic contexts.

References

1. Alligood, K.T., Sauer, T., Yorke, J.A.: Chaos. An Introduction to Dynamical Systems. Springer, New York (1996)
2. Andrecut, M.: Logistic map as a random number generator. Int. J. Mod. Phys. B **12**(9), 921–930 (1998)
3. Barkley Rosser, J.: From Catastrophe to Chaos: A General Theory of Economic Discontinuities. Springer, New York (2000)
4. Baumol, W.J., Benhabib, J.: Chaos: significance, mechanism, and economic applications. J. Econ. Perspect. **3**(1), 77–105 (1981)
5. Benhabib, J.: On cycles and chaos in economics. Stud. Nonlinear Dyn. Econom. **1**(1), 1–4 (1996)
6. Benhabib, J., Day, R.H.: Rational choice and erratic behaviour. Rev. Econ. Stud. **48**(3), 459–471 (1981)
7. Bullard, J., Butler, A.: Nonlinearity and chaos in economic models: implications for policy decisions. Econ. J. **103**(419), 849–867 (1993)
8. Day, R.H.: Irregular growth cycles. Am. Econ. Rev. **72**(3), 406–414 (1982)

9. Day, R.H.: The emergence of chaos from classical economic growth. Q. J. Econ. **98**(2), 201–213 (1983)
10. Day, R.H.: Complex economic dynamics: obvious in history, generic in theory, elusive in data. J. Appl. Econom. **7**(Suppl.), 9–23 (1992). Special Issue on Nonlinear Dynamics and Econometrics
11. Emmert-Streib, F., Behmer, M.: Optimization procedure for predicting non-linear time series based on a non-Gaussian noise model. In: Gelbukh, A., Fernando, A. (eds.) MICAI 2007: Advances in Artificial Intelligence. Springer, Heidelberg (2007)
12. Faggini, M., Parziale, A.: The failure of economic theory. Lessons from chaos theory. Mod. Econ. **3**, 1–10 (2012)
13. Farmer, R.: Deficits and cycles. J. Econ. Theory **40**(1), 77–88 (1986)
14. Fidanova, S., Roeva, O., Mucherino, A., Kapanova, C.: InterCriteria analysis of ant algorithm with environment change for GPS surveying problem. In: Dichev, C., Agre, G. (eds.) Artificial Intelligence: Methodology, Systems, and Applications: AIMSA 2016 Proceedings. Springer, Cham (2016)
15. Grandmont, J.M.: On endogenous competitive business cycles. Econometrica **53**(5), 995–1045 (1985)
16. Grandmont, J.M.: Stabilizing competitive business cycle. J. Econ. Theory **40**(1), 57–76 (1986)
17. Grandmont, J.M.: Nonlinear difference equations, bifurcations and chaos: an introduction. Res. Econ. **62**, 122–177 (2008)
18. Holyst, J.A.: How to control a chaotic economy. J. Evol. Econ. **6**(1), 31–42 (1996)
19. Holyst, J.A., Urbanowicz, K.: Chaos control in economical model by time delayed feedback method. Physica A **287**(3–4), 587–598 (2000)
20. Kaas, L.: Stabilizing Chaos in a Dynamic Macroeconomic Model. J. Econ. Behav. Organ. **33** (3–4), 313–332 (1998)
21. Koda, M.: Chaos search in fourier amplitude sensitivity test. J. Inf. Commun. Technol. **11**, 1–16 (2012)
22. Kopel, M.: Improving the performance of an economic system: controlling chaos. J. Evol. Econ. **7**(3), 269–289 (1997)
23. May, R.M.: Simple mathematical models with very complicated dynamics. Nature **261**, 459–467 (1976)
24. Medio, A.: Teoria Non-Lineare del Ciclo Economico. Il Mulino, Bologna (1979)
25. Nishimura, K., Sorger, G.: Optimal Cycles and Chaos: A Survey. Stud. Nonlinear Dyn. Econom. **1**(1), 11–28 (1996)
26. Persohn, K.J., Povinelli, R.J.: Analyzing logistic map pseudorandom number generators for periodicity induced by finite precision floating-point representation. Chaos Solitons Fractals **45**(3), 238–245 (2012)
27. Thom, R.: Stabilité Structurelle et Morphogénèse. W.A. Benjamin Inc., Massachusetts (1972). English Edition: Structural Stability and Morphogenesis. Westwiew Press, Reading, Massachusetts (1975)
28. Tzouveli, P.K., Ntalianis, K.S., Kollias, S.D.: Chaotic encryption driven watermarking of human video objects based on Hu moments. In: Maglogiannis, I., Karpuozis, K., Bramer, M. (eds.) Artificial Intelligence Applications and Innovations. AIAI 2006 Proceedings. Springer, Boston (2006)

Optimal Noise Manipulation in Asymmetric Tournament

Zhiqiang Dong[1]($\boxtimes$) and Zijun Luo[2]

[1] Key Lab for Behavioral Economic Science and Technology, South China
Normal University, Guangzhou 510006, China
dongzhiqiang@m.scnu.edu.cn
[2] Department of Economics and International Business, Sam Houston State
University, Huntsville, TX 77341-2118, USA

Abstract. We fill a gap in the literature of asymmetric tournament by allowing
the principal to optimally alter noise in relative performance evaluation, such
that the observed performance of each agent is less or more dependent of ability
and effort. We show that there exists an optimal noise level from the principal's
standpoint of expected profit maximization. It is shown that this optimal noise
level is higher than what would induce the highest efforts from the two agents.

Keywords: Asymmetric tournament · Noise manipulation · Incentive contract

1 Introduction

In the tournament literature pioneered by Lazear and Rosen (1981), the noise that affects
agents' observed performance is assumed to be exogenous. In theory, the greater the
noise, the less impact efforts will have on performance, and the lower the agents' efforts
will be. As a result, in order to achieve the same level of output as a noiseless environment, a larger reward is necessary to incentivize the agents when noise is notable. This
idea is considered as a basic principle in promotion tournaments (Lazear 1998, p. 232).

However, Lazear's argument is only a partial story. Although a capricious principal
will lead to loss of efficiency when agents are homogenous or similar, it is not necessarily the case with sufficient heterogeneity between agents. The presence of luck,
i.e., noise, may stimulate the agents to work harder. Intuitively, with some luck, the
weaker agent's expected probability to win may be higher, which can lead an increase
in effort. The stronger agent, with an *ex ante* expectation to win, may respond by also
increasing her effort. In such a case, the unpredictable behavior of the principal may not
be inefficient. The principal is able to strategically control "luck" when the agents are
known to be heterogeneous.

In this paper, we develop a principal-agent model to incorporate the aforementioned
intuition. The analysis of risk or noise is largely missing in the tournament and contest
literature (Lazear and Rosen 1981; Green and Stokey 1983; Rosen 1986; Bhattacharya
and Guasch 1988; Yun 1997; Moldovanu and Sela 2001; Nti 2004) with the exceptions
of O'keeffe et al. (1984), Skaperdas and Gan (1995), Cowen and Glazer (1996), Hvide
(2002), Kräkel and Sliwka (2004), Gilpatric (2009), Wang (2010), Kwon (2012),
Kwon (2013), and Kellner (2015), to name a few.

© Springer Nature Switzerland AG 2019
E. Bucciarelli et al. (Eds.): DCAI 2018, AISC 805, pp. 28–35, 2019.
https://doi.org/10.1007/978-3-319-99698-1_4

Our paper is closely related to Wang (2010) and Kwon (2012). In a contest setting with the Tullock contest success function (Tullock 1980), Wang (2010) allows the designer to choose the accuracy level. The paper finds that the optimal accuracy level is negatively related to the difference in abilities, i.e., higher difference results in lower accuracy level. On the other hand, in a dynamic (2-stage) tournament setting, Kwon (2012) finds that the optimal noise chosen by the designer is non-zero unless for symmetrical agents. In our model, agents compete in a single stage. While we also find that non-zero noise to be optimum, our results indicate a positive relationship between difference in abilities and the noise level.

2 Model Setup

Consider a typical setting of tournament in which a principal employs two asymmetric agents and implements relative performance evaluation. The output of agent i, denoted as y_i for $i = 1, 2$, is given by

$$y_i = K_i + a_i + \epsilon_i, \tag{1}$$

where K_i and a_i are ability and effort level of agent i, which is unobservable by the principal and the other agent, and ϵ_i is noise naturally occurred in the production process and is independently and identically distributed. Without loss of generality, we assume $\epsilon_i \sim N(0, \frac{1}{2})$ throughout this paper. We follow the literature in assuming that abilities and efforts are perfect substitutes.

To incorporate noise manipulation by the principal, we assume that the observed outcome of agent i in relative performance evaluation is given by

$$\hat{y}_i = K_i + a_i + \lambda \epsilon_i \tag{2}$$

where $\hat{y}_i$ is the observed outcome and $\lambda > 0$ is the coefficient of noise manipulation. The principal can manipulate λ in order to change the noise in production. All players can observe $\hat{y}_i$ but not a_i or y_i.

The risk-neutral principal's goal is to maximize expected profit given by

$$V = E[y_1 + y_2] - 2w - R = a_1 + a_2 - 2w - R \tag{3}$$

where w and R are, respectively, competitive wage and reward to be paid to the agents and determined endogenous by the principal in the model. The reward is only given to the agents with the higher value of $\hat{y}_i$.

The timeline of the game is as follows. In the first stage, the principal chooses the coefficient of noise manipulation, λ. In the second stage, the principal designs the contract (w and R). In the third and last stage, the two agents independently choose a_i. The game is solved using backward induction.

3 Results

We first solve for the optimal decision of the agents in the last stage of the game. From Eq. (2), the winning probability of agent 1 is given by

$$
\begin{aligned}
&\Pr[K_1 + a_1 + \lambda\epsilon_1 > K_2 + a_2 + \lambda\epsilon_2] \\
&= \Pr[\lambda(\epsilon_1 - \epsilon_2) > a_2 - a_1 - k] \\
&= 1 - G(a_2 - a_1 - k)
\end{aligned}
\tag{4}
$$

where $G(.)$ is the cumulative distribution function (*CDF*) of random variable $\lambda(\epsilon_1 - \epsilon_2) \sim N(0, \lambda^2)$, and $k = K_1 - K_2 > 0$ captures the difference in the abilities of the two agents with $K_i > 0$. In this specification, $k > 0$ indicates that agent 1 is of higher ability and $k < 0$ otherwise. For risk-neutral agents, we assume their expected utilities, u_i, to be expressed by the following quadratic functions

$$
\begin{cases}
u_1 = w + [1 - G(a_2 - a_1 - k)]R - \frac{a_1^2}{2} \\
u_2 = w + G(a_2 - a_1 - k)R - \frac{a_2^2}{2}
\end{cases}
\tag{5}
$$

where $\frac{a_i^2}{2}$ is the disutility of effort for agent i. The first order conditions (FOCs) for Eq. (5) are

$$
\begin{cases}
g(a_2 - a_1 - k)R = a_1 \\
g(a_2 - a_1 - k)R = a_2
\end{cases}
\tag{6}
$$

where $g(.)$ is the probability density function (*pdf*) of $\lambda(\epsilon_1 - \epsilon_2)$. We thus obtain the solutions for a_i from Eq. (6) as

$$
a_i^* = a^* = g(-k)R, i = 1, 2.
\tag{7}
$$

Substituting in the *pdf* of normal distribution and taking the derivative with respect to (w.r.t.) λ, we obtain

$$
\frac{\partial a_i^*}{\partial \lambda} = \frac{k^2 - \lambda^2}{\lambda^4 \sqrt{2\pi}} \exp\left(-\frac{k^2}{2\lambda^2}\right) R.
\tag{8}
$$

We can establish the following lemma according to this derivative.

Lemma 1. When agents are homogeneous ($k = 0$) and the reward is positive ($R > 0$), the optimal efforts (a_i^*) of the agents decrease as noise (λ) increases. However, when agents are heterogeneous ($k > 0$), the optimal efforts increase as noise increases if noises are low ($\lambda \in (0, k]$) and decrease if noises are high ($\lambda > k$).

Proof: The result follows immediate from Eq. (8).   ∎

Lemma 1 implies that the principal can choose $\lambda = k$ to maximize effort of each agent. However, it will be shown later that the optimal noise in term of the principal's profit maximization falls in the interval $\lambda > k$. In other words, the profit-maximizing noise level exceeds that of effort-maximizing.

We now investigate the second stage of the game. Substituting Eq. (7) into Eq. (3) and taking into account both agents individual rationality constraints (IRs) as well as the limited liability constraint (LL), we rewrite the principal's expected profit maximization problem into

$$\max_{\{w,R\}} V = [2g(-k) - 1]R - 2w$$

$$\text{s.t. } w + [1 - G(-k)]R - \tfrac{a^2}{2} \geq 0 \, (\text{IR1}) \tag{9}$$

$$w + G(-k)R - \tfrac{a^2}{2} \geq 0 \, (\text{IR2})$$

$$w \geq 0 \, (\text{LL})$$

It is straightforward to verify that (LL) and (IR2) are binding while (IR1) is not. As a result, we obtain $w = 0$ and rewrite the problem, after substituting in the solution of a^*, as

$$\max_{\{R\}} V = [2g(-k) - 1]R$$

$$\text{s.t. } G(-k)R - \tfrac{[g(-k)R]^2}{2} \geq 0 \tag{10}$$

Solving for R, we obtain

$$R^* = \begin{cases} 0 & \text{if } g(-k) \leq \tfrac{1}{2} \\ \dfrac{2G(-k)}{g(-k)^2} & \text{if } g(-k) > \tfrac{1}{2} \end{cases} \tag{11}$$

Since $a^* = 0$ and $V^* = 0$ when $R^* = 0$, we focus only on the case where $g(-\Delta k) > \tfrac{1}{2}$. Under this condition, we are able to establish the following lemma.

Lemma 2. There exists a close feasible set for the principal to extract efforts from agents in the tournament. This close feasible set requires $k < \bar{k} \equiv \lambda \sqrt{\log[2(\pi\lambda^2)^{-1}]}$ with $\lambda^{max} = \sqrt{2/\pi}$ and $\bar{k}^{max} = \sqrt{2/(\pi e)}$.

Proof: If $g(-k) < \tfrac{1}{2}$, the principal prefers $R^* = 0$ (hence $a^* = 0$ and $V^* = 0$) rather than any $R > 0$. The principal prefers $R > 0$ and extracts a positive effort $a^* > 0$ (in the feasible set) if and only if $g(-k) > \tfrac{1}{2}$. Plugging the normal distribution *pdf* into the condition shows the first part of Lemma 2 after simplification. For λ^{max}, notice that the value of $\bar{k}$ is constrained by $\log\left[2\left(\pi\lambda^2\right)^{-1}\right] \geq 0$ which in turn constraints λ and gives $\lambda \leq \lambda^{max} \equiv \sqrt{2/\pi}$. For $\bar{k}^{max}$, we take the first order derivative of $\bar{k}$ w.r.t. λ and obtain $\bar{k}_\lambda = \frac{\log[2(\pi\lambda)^{-2}] - 1}{\sqrt{\log[2(\pi\lambda)^{-2}]}}.$ [1] Setting the first order derivative to 0, we obtain $\lambda^{max} = \sqrt{2/\pi}$ and

---

[1] Note that the second derivative, $\bar{k}_{\lambda\lambda} = -\tfrac{2}{3}\left(\frac{1 + \log[2/(\pi\lambda^2)]}{\lambda \log[2/(\pi\lambda^2)]}\right)$, is negative. This guarantees the strict concavity of $\bar{k}$ as a function of λ.

hence $\bar{k}^{max} = \sqrt{2/(\pi e)}$. ∎

When $g(-k) > \frac{1}{2}$ is satisfied, we have

$$R^* = \frac{2G(-k)}{[g(-k)]^2};$$
$$a^* = \frac{2G(-k)}{g(-k)};$$
$$V^* = \frac{4G(-k)}{g(-k)} - \frac{2G(-k)}{[g(-k)]^2}.$$

We can now proceed to the first stage of the game and solve for the optimal value of λ. Taking the first derivative of Eq. (10) w.r.t. λ, we obtain

$$\frac{\partial V}{\partial \lambda} = \frac{\partial}{\partial \lambda}\left(\frac{4G(-k)}{g(-k)}\right) - \frac{\partial}{\partial \lambda}\left(\frac{2G(-k)}{[g(-k)]^2}\right) \equiv MR - MC \tag{11}$$

where

$$MR = \frac{1}{\lambda^2}\left[4k\lambda - 4(k^2 - \lambda^2)\sqrt{2\pi}\operatorname{Exp}\left(\frac{k^2}{2\lambda^2}\right)G(-k)\right] \tag{12}$$

$$MC = \frac{1}{\lambda}\left[2k\lambda\sqrt{2\pi}\operatorname{Exp}\left(\frac{k^2}{2\lambda^2}\right) - 8(k^2 - \lambda^2)\pi\operatorname{Exp}\left(\frac{k^2}{\lambda^2}\right)G(-k)\right] \tag{13}$$

From Eq. (11), we can establish the following proposition

Proposition 1. When $\lambda \in [k, \lambda^{max}]$ and parameters are in the principal's feasible set, there exists a unique λ^* that maximizes the expected profit of the principal.

Proof: First, it can be verified that when $\lambda \in [k, \lambda^{max}]$, marginal revenue (MR) and marginal cost (MC) are both monotonic and increase in λ. When $\lambda = k$, we have $MR_{\lambda=k} = 4$ and $MC_{\lambda=k} = 2k\sqrt{2\pi e} < 2\bar{k}^{max}\sqrt{2\pi e} = 4$. As a result, when $\lambda = k$, we have $MR_{\lambda=k} > MC_{\lambda=k}$. On the other end, when $\lambda = \lambda^{max}$, $k = 0$. Substituting λ^{max} and $k = 0$ into Eq. (11), we have $MR_{\lambda=\lambda^{max}} - MC_{\lambda=\lambda^{max}} = -4G(0)\sqrt{2\pi} = -2\sqrt{2\pi} < 0$. As a result, $MR_{\lambda=\lambda^{max}} > MC_{\lambda=\lambda^{max}}$. This proves that there exists a single crossing of MR and MC in $[k, \lambda^{max}]$. ∎

Proposition 1 indicates that the effort-maximizing risk level $\lambda = k$ does not maximize the principal's profit, since the principal needs to provide extra incentive to the agents for the extra efforts due to the individuals rationality and limited liability constraints. Furthermore, a noiseless environment also undermines the principal's profit. The optimal, profit-maximizing, level of risk chosen by the principal exceeds the effort-maximizing level. In such case, the principal injects "luck" into the agents' outputs and the agents exert lower efforts but the principal reaches the maximum profit.

4   Simulation and Numerical Examples

This section provides some simulation and numerical results of the Model. In Fig. 1, the shaded area between $\bar{k}$ (the inverted-U shape curves) and the horizontal axis is the principal's feasible set, where $R > 0$ is preferred to $R = 0$ for the principal. The principal gets zero profit at any point on curve $\bar{k}$, and gets positive (negative) profit in the area below (above) curve $\bar{k}$. The path $\lambda^*(k)$ (solid line going southwest to northeast) depicts the optimal choices of the principal given different values of k, as outlined in Proposition 1. Figure 1 shows that an optimal λ with positive profit occurs only when $\lambda > k$ (below the 45° line), and that λ is increasing in k. The intersection of path λ^* and the 45° line is $\bar{k}^{max} = \sqrt{2/(\pi e)}$.

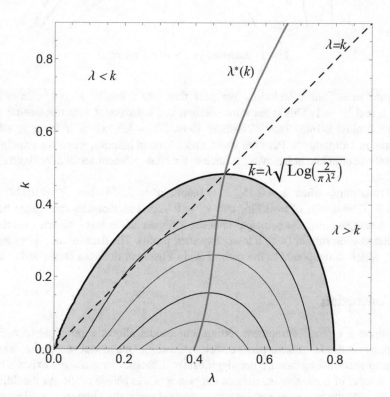

Fig. 1. Feasible set of noise manipulation and the optimal noise pattern

Figure 2 depicts the expected profit of the principal when $K_2 = 0$ and hence $k = K_1$. We examine four values of k: 0, 0.15, 0.3, and 0.5, respectively. A higher curve is corresponding to a lower value of k, i.e., less heterogeneity between the agents. Figure 2 shows that the larger the heterogeneity between agents, the lower the principal's expected profit. The expected profit even turns into negative when $k > k^{max} (= \sqrt{2/\pi e} \approx 0.484)$. This is shown by the curve $V(k = 0.5)$.

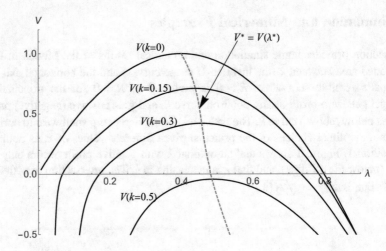

Fig. 2. Expected profits of the principal

From numerical calculation, we find that when $k = 0$, $\lambda^* = 1/\sqrt{2\pi} \approx 0.399$, $a_i^* = 1$, and $V^* = 1$. This is the same solution to a tournament with risk-neutral agents without limited liability (see Wolfstetter 1999, Chap. 12), which is the most efficient outcome in tournaments. Put differently, under limited liability, when the principal can optimally manipulate noise, we can achieve the same efficient result as without limited liability.

Furthermore, when $k = 0.15$, $\lambda^* = 0.438$, $a^* = 0.854$, and $V^* = 0.712$; when $k = 0.3$, $\lambda^* = 0.457$, $a^* = 0.726$, and $V^* = 0.420$. Therefore, as the agents become more heterogeneous, the principal chooses a higher noise level, which distorted the production even further (with a lower expected profit). The dashed line shows the path of V^*, which corresponds to the path of λ^* in Fig. 1, at different levels of k.

5 Conclusion

Built upon a typical tournament setting, our model allows a principal to optimally manipulate noise in relative performance evaluation. Our model shows that when the two heterogeneous agents are not significantly different, there always exists a unique optimal level of noise that maximizes the principal's expected profit. As the difference in abilities of the two agents widens, the principal optimally chooses a noisier scheme, resulting in lower efforts by the two agents and a lower overall profit for the principal. Our analysis helps to explain why managers tend to choose workers with similar abilities and why organizations conducting similar business may have very different performance evaluation standards with different accuracy due to heterogeneities among workers. Our conjecture is that results from the current model will generally hold with more than two agents.

References

Bhattacharya, S., Guasch, J.L.: Heterogeneity, tournaments, and hierarchies. J. Polit. Econ. **96**, 867–881 (1988)

Cowen, T., Glazer, A.: More monitoring can induce less effort. J. Econ. Behav. Organ. **30**, 113–123 (1996)

Gilpatric, S.M.: Risk taking in contests and the role of carrots and sticks. Econ. Inq. **47**, 266–277 (2009)

Green, J.R., Stokey, N.L.: A comparison of tournaments and contracts. J. Polit. Econ. **91**, 340–364 (1983)

Hvide, H.K.: Tournament rewards and risk taking. J. Labor Econ. **20**, 877–898 (2002)

Kellner, C.: Tournaments as a response to ambiguity aversion in incentive contracts. J. Econ. Theory **59**, 627–655 (2015)

Kräkel, M., Sliwka, D.: Risk taking in asymmetric tournaments. Ger. Econ. Rev. **5**, 103–116 (2004)

Kwon, I.: Noisy and subjective performance measure in promotion tournaments. Seoul J. Econ. **25**, 207–221 (2012)

Kwon, I.: Risk-taking in subjective promotion tournaments. Appl. Econ. Lett. **20**, 1238–1243 (2013)

Lazear, E.P.: Personnel Economics for Managers. Wiley, New York (1998)

Lazear, E.P., Rosen, S.: Rank-order tournament as optimum labor contracts. J. Polit. Econ. **89**, 841–864 (1981)

Moldovanu, B., Sela, A.: The optimal allocation of prizes in contests. Am. Econ. Rev. **91**, 542–558 (2001)

Nti, K.O.: Maximum efforts in contests with asymmetric valuations. Eur. J. Polit. Econ. **20**, 1059–1066 (2004)

O'keeffe, M., Viscusi, W.K., Zeckhauser, R.J.: Economic contests: comparative reward schemes. J. Labor Econ. **2**, 27–56 (1984)

Rosen, S.: Prizes and incentives in elimination tournaments. Am. Econ. Rev. **76**, 701–715 (1986)

Skaperdas, S., Gan, L.: Risk aversion in contests. Econ. J. **105**, 951–962 (1995)

Tullock, G.: Efficient rent seeking. In: Buchanan, J., Tollison R., Tullock, G. (eds.) Toward a Theory of the Rent-Seeking Society, pp. 97–112. Texas A&M University Press, College Station (1980)

Wang, Z.: The optimal accuracy level in asymmetric contests. B.E. J. Theor. Econ. **10** (2010). Article 13

Wolfstetter, E.: Topics in Microeconomics: Industrial Organization, Auctions, and Incentives. Cambridge University Press, Cambridge (1999)

Yun, J.: On the efficiency of the rank-order contract under moral hazard and adverse selection. J. Labor Econ. **15**, 466–494 (1997)

Decision-Making Process Underlying Travel Behavior and Its Incorporation in Applied Travel Models

Peter Vovsha[✉]

WSP, 1Penn Plaza, 2nd Floor, New York, NY 10119, USA
Peter.Vovsha@WSP.com

Abstract. The paper provides a broad overview of the state of the art and practice in travel modeling in its relation to individual travel behavior. It describes how different travel decision making paradigms led to different generations of applied travel models in practice – from aggregate models to disaggregate trip-base models, then to tour-based models, then to activity-based models, and finally to agent-based models. The paper shows how these different modeling approaches can be effectively generalized in one framework where different model structures correspond to different basic assumptions on the decision-making process. The focus of the paper is on three key underlying behavioral aspects: (1) how different dimensions of travel and associated individual choices are sequenced and integrated, (2) how the real-world constraints on different travel dimensions are represented, (3) what are the behavioral factors and associated mathematical and statistical models applied for modeling each decision-making step. The paper analyzes the main challenges associated with understanding and modeling travel behavior and outlines avenues for future research.

Keywords: Travel behavior · Agent-based modeling
Decision-making process

1 Individual Travel Behavior and Applied Travel Models

1.1 Travel Demand Model Types

Historically, applied travel models (i.e. travel models adopted by a public agency as the modeling tool for planning and project evaluation) started with a purely aggregate approach mimicking kinematic models in physics like a gravity model of spatial trip distribution [1]. This led to the aggregate 4-Step modeling paradigm further extended to aggregate Tour-Based models [2]. These models have a very little behavioral realism embedded in the model structure. In time, the recognition that the main factors affecting travel behavior can be better described and modeled at the individual level resulted in a new direction that led to disaggregate Activity-Based Models (ABMs) [3, 4], and finally to Agent-Based Models (AgBMs) implemented in a microsimulation fashion [5, 6].

Despite the disaggregate nature of ABMs and AgBMs they still have to serve the main purpose of generating a credible aggregate travel forecast for transportation

© Springer Nature Switzerland AG 2019
E. Bucciarelli et al. (Eds.): DCAI 2018, AISC 805, pp. 36–48, 2019.
https://doi.org/10.1007/978-3-319-99698-1_5

planning. In this regard these models still should satisfy important criteria that are not automatically ensured by advanced disaggregate model structures:

- Travel model should match independent aggregate targets such as ground traffic counts or transit ridership.
- Travel models are primarily used for comparisons of future scenarios or transportation infrastructure alternatives. Transparency and logical sensitivity to policies proved to be an aspect difficult to address. In this regard, such feature as aggregate negative cross-elasticity (if something is improved in transportation network all competing options should lose users) becomes mandatory. This feature is ensured in simplified aggregate models but is not automatically held in ABMs or AgBMs [1].

1.2   Applied Travel Model and Its Dimensions

Any applied travel model of the four types described above takes the following input and produce the following output:

- Input: socio-economic data by zone (population employment) and transportation network
- Output: predicted trips (by origin & destination zones, trip purpose (activity type at origin & destination), departure time, mode, and network route) and network performance measures such as trip Level of Service (LOS) in terms of time, cost, convenience, reliability and transportation facility loading.

A regional travel model represents a typical case to advocate AgBM approach due to its complexity in terms of number of agents and dimensions of their behavior as well as multiple layers of interactions between the agents. This recognition led to the interest in individual travel behavior, its dimensions, and underlying decision-making process [2].

2   Behavioral Decision Processes Associated with Travel

2.1   Complexity of Individual Travel Behavior and Its Dimensions

Individual Daily Activity Pattern and schedule in general can be described as a sequence of trips and out-of-home activities undertaken by the individual in time and space. Spatial and temporal details contribute to the horrendous dimensionality of DAP that cannot be formulated as a single choice model [6]. Another complex aspect of individual travel behavior is that it spans multiple temporal scales (from long term of 20+ years to short term of a day or hour) for decision-making with a reliance on gradual learning. While the ultimate objective of an applied travel model is to predict travel for a day the associated decision making and learning are not bound to a real-time day framework. Interestingly, however, the entire regional model of travel behavior of any of the four types can be described in a compact way through a limited number of key dimensions although its practical implementation will require a system of interconnected models mimicking the associated decision-making steps.

2.2   Microfoundations and Micro-Macro Relationships

Travel behavior and associated decision-making processes are inherently individual. However, there are three important reasons why a travel model still needs to address aggregate macro constraints and factors that calls to an iterative equilibration structure of all travel models:

- Essential demand-supply equilibration due to aggregate capacity constraints. There are two major aggregate capacity constraints imposed on individual travel behavior. The first one relates to a constrained activity supply that is expressed in a competition for jobs and student enrollment as well as in time-varying accessibility measures (opening and closing hours of establishments) that affect discretionary trips [6]. The second one relates to a constrained transportation network capacity that results in highway network congestion effects, transit network crowding effects, and parking constraints [7].
- Learning from others and "contagion" effects, for example, with respect to individual mobility attributes and modality preferences affected by aggregate modal split [8].
- Effective using of aggregate proportions as basis for modeling (drawing) individual choices (so-called "data-driven" approach). It is especially appealing for spatial trip distribution where taking advantage of "big data" and necessity to validate the applied travel model call for addressing this micro-macro relationship.

2.3   Group Decision Making

Most of the travel decisions are made at two levels – household and person where persons strongly interact within the household (which has been reflected in the advanced applied ABMs and AgBMs) and there are also inter-household interactions that are however less explored. Joint travel directly relates to important transportation policies and it has been largely mismodeled in the past by ignoring its specifics in aggregate 4-step models. Car occupancy is not an individual mode choice but rather an example of a complex time-space synchronization between several persons. The following examples of group decision making mechanisms were successfully incorporated in applied travel models [9–11]:

- Coordinated choice of workplaces for multiple workers that can be further integrated with residential location choice.
- Coordinated schedules of household members including usual work schedules and family time spent at home.
- Coordinated daily pattern types (person day-offs).
- Joint activities and associated travel (in particular, fully-joint tours).
- Joint travel arrangements and escorting including escorting children to school, escorting children to discretionary activities, and joint commuting.

2.4   Evolution of Travel Models

Historical overview, reasons for moving from generation to generation and associated assumptions on decision-making process are summarized in Table 1. The major shifts relate to the objects and subjects of the model that became more and more disaggregate on both activity demand and supply sides. Additionally, each new modeling paradigm tried to address a set of new aspects of internal model consistency in time and space [1–3, 5] including the way how the interactions between the modeled objects and subjects are portrayed [6–10].

Table 1. Evolution of travel models and underlying decision-making paradigms.

Aspect	4-step	Tour-based	ABM	AgBMs
Main object	Trips by segment	Closed trip chain (tour) by segment	Individual daily trip chain	Individual daily activity-trip chain
Main subject of activity-travel demand	Zones with population that generates trips		Individuals that participate in out-of-home activities and implement trips	
Main subject of activity supply	Zones with land-uses that attracts trips			Individual establishments that attract trips
Direct interaction between agents	Ignored		Intra-household	Intra-household & inter-household
Individual learning & adaptation	Ignored			Dynamically updated model parameters
Trip end constraints	Zonal trip ends match production & attraction constraints for each segment			
HB and NHB trips match	Ignored	Zonal production and attraction for Home-based (HB) and Non-home-based (NHB) trips match		
Trip mode consistency	Ignored	Trip modes are consistent across trip chains (tours)		
Schedule consistency	Ignored		Activities and tours for the person do not overlap in time	
Time-space consistency	Ignored			Activities and tours feasible in time & space
Household car allocation	Ignored			Consistent with mode choice
Parking constraint	Ignored		Zonal car trip ends are consistent with parking capacity	

3 Closer Look at Each Modeling Paradigm

3.1 Key Behavioral Aspects for Travel Model Construction

In the subsequent sections, the key aspects of travel behavior will be summarized in the context of travel model construction. For each of the four major modeling paradigms the following three aspects are analyzed:

1. How different dimensions of travel and associated individual choices are sequenced and integrated.
2. How the real-world constraints on different travel dimensions are represented.
3. What are the behavioral factors and associated models applied at each decision-making step.

3.2 Four-Step Model

Main structural elements of a typical 4-step model are summarized in Table 2. The key subjects of the model are zones where a pair of Origin and Destination zones (OD pair) defines the geography of trips as the main modeled objects. The essence of the model is to predict number of trips for each OD pair by mode, Time of Day (TOD), and network route [1]. Trips are frequently segmented by trip purpose or other attributes.

Table 2. Sequence of travel choices for 4-step model

Step	Sub-model	Loops over objects
1 = Trip generation	1.1 = Trip production	Segment, origin
	1.2 = Trip attraction	Segment, destination
2 = Trip distribution	2.1 = Seed or impedance matrix	Segment, OD pair
	2.2 = Balancing by origins	Segment, origin
	2.3 = Balancing by destinations	Segment, destination
	2.4 = TOD slicing	Segment, OD pair
3 = Mode choice	3.1 = Mode-specific generalized cost	Segment, TOD period, OD pair
	3.2 = Mode trip tables	Segment, TOD period, OD pair
4 = Assignment	4.1 = Assignable trip tables	TOD period, mode, OD pair
	4.2 = Route choice and network loading	TOD period, mode, OD pair
	4.3 = Level of service skimming	TOD period, mode, OD pair

Decision-making paradigm underlying a 4-step model can be summarized as follows:

- Trips are generated independently of each other as flows from zone to zone. As the result, the model is highly inconsistent internally [2, 3].

- For each generated trip the underlying sequence of choices is Destination → Time of Day (TOD) → Mode → Network route [1].
- Segmentation is very limited and includes only trip purpose and a few household attributes like income group and car ownership [6].

3.3   Tour-Based Model

Main structural elements of a typical Tour-Based model are summarized in Table 3. The principal difference compared to the 4-step model is that the main modeling object was replaced with a trip chain (tour) that opens a way to make trip-level choices more consistent and account for additional constraints [2–4].

Table 3. Sequence of travel choices for tour-based model

Step	Sub-model	Loops over objects
1 = Tour generation	1.1 = Tour production	Segment, origin
	1.2 = Tour attraction (primary destination)	Segment, destination
	1.3 = Intermediate stop attraction	Segment, destination
2 = Tour distribution	2.1 = Seed or impedance matrix	Segment, OD pair
	2.2 = Balancing by origins	Segment, origin
	2.3 = Balancing by destinations	Segment, destination
3 = Tour TOD	3.1 = TOD-specific impedance	Segment, OD pair
	3.2 = Tour TOD (outbound, inbound)	Segment, OD pair
4 = Stop insertion	4.1 = Stop frequency	Segment, OD pair, TOD, direction
	4.2 = Stop location	Segment, OD pair, TOD, direction, frequency
5 = Mode choice	5.1 = Mode-specific generalized cost	Segment, OD pair, TOD, direction, frequency
	5.2 = Mode trip tables	Segment, OD pair, TOD, direction, frequency
6 = Assignment	6.1 = Assignable trip tables	TOD period, mode, OD pair
	6.2 = Route choice and network loading	TOD period, mode, OD pair
	6.3 = Level of service skimming	TOD period, mode, OD pair

Decision-making paradigm underlying a tour-based model can be summarized as follows:

- Trips are generated within tours and chained in a way that trip destination, TOD, and mode choices are coordinated within the tour.
- Tours are generated independently of each other as flows of round trips from zone to zone which makes the model highly inconsistent internally beyond the tour level.

- For each generated tour the underlying sequence of choices is Primary destination → TOD by direction (outbound, inbound) → Intermediate stop frequency and location → Tour mode (frequently assumed the same for each trip on the tour) → Network route for each trip.
- Segmentation is very limited and includes tour/trip purpose and a few household attributes like income group and car ownership; this is similar to a 4-step model.

3.4 Activity-Based Model (ABM)

Main structural elements of a typical ABM are summarized in Table 4. ABM represents a principal shift towards individual persons and households as main modeled subjects that allows for much richer segmentation and opens many additional lines for consistency and constraining travel choices within a complete individual daily activity pattern [3, 6, 10, 11].

Table 4. Sequence of travel choices for activity-based model

Step	Sub-model	Loops over objects
1 = Long-term choices	1.1 = Usual workplace and school	Workers and students
	1.2 = Household car ownership	Households
2 = Daily activity pattern	2.1 = Daily pattern type	Households, persons
	2.2 = Joint tour frequency & participation	Households, persons
	2.3 = Individual mandatory tours	Workers and students
	2.4 = Individual non-mandatory tours	Households, persons
3 = Tour details	3.1 = Non-mandatory tour destination	Households, persons, non-mandatory tours
	3.2 = Tour TOD (outbound, inbound)	Households, persons, tours
	3.3 = Tour mode choice	Households, persons, tours
	3.4 = Stop frequency	Households, persons, tours
4 = Trip details	4.1 = Stop location	Households, persons, tours, stops
	4.2 = Trip mode choice	Households, persons, tours, trips
	4.3 = Trip departure time	Households, persons, tours, trips
5 = Assignment	5.1 = Assignable trip tables	TOD period, mode, OD pair
	5.2 = Route choice and network loading	TOD period, mode, OD pair
	5.3 = Level of service skimming	TOD period, mode, OD pair

Decision-making paradigm underlying ABM can be summarized as follows:

- Trips are generated within tours and chained. Trip destination, TOD, and mode choices are coordinated within the tour.
- Tours are generated for each individual within daily patterns without an overlap in time.
- For each generated tour the underlying sequence of choices is Primary destination → TOD by direction (outbound, inbound) → Tour mode → Stop frequency.
- Subsequently, for each generated trip the underlying sequence of choices is Stop location → Trip mode → Trip departure time → Network route.
- Segmentation is unlimited and includes tour/trip purpose and multiple household & person attributes like income group, car ownership, household size and composition, person type, age, and gender.
- Time-space constraints are not fully accounted.

3.5 Agent-Based Model (AgBM)

Main structural elements of a typical AgBM are summarized in Table 5. AgBM is a natural extension of ABM with taking a full advantage of individual microsimulation and specifically ensuring that time-space constraints are fully accounted [2, 12]. Another key difference from ABM is a certain level of intelligence, learning, and adaptation of the modeled subjects that is only pertinent to AgBM.

Table 5. Sequence of travel choices for agent-based model

Step	Sub-model	Loops over objects
1 = Long-term choices	1.1 = Usual workplace and school	Workers and students
	1.2 = Household mobility attributes	Households
2 = Daily activity pattern	2.1 = Daily pattern type	Households, persons
	2.2 = Joint activities & escorting arrangements	Households, persons
	2.3 = Individual mandatory activities	Workers and students
	2.4 = Individual non-mandatory activities	Households, persons
3 = Tour formation	3.1 = Allocation of activities to day segments	Households, persons, activities
	3.2 = Activity sequencing	Households, persons, day segments, activities
	3.3 = Activity locations and tour starts/ends	Households, persons, day segments, activities
4 = Tour & trip details	4.1 = Tour TOD	Households, persons, tours
	4.2 = Combinatorial tour/trip mode choice	Households, persons, tours, trips

(*continued*)

Table 5. (*continued*)

Step	Sub-model	Loops over objects
	4.3 = Within-tour time allocation & trip departure	Households, persons, tours, trips
	4.4 = Car allocation and use (routing) details	Households, persons, tours, trips, cars
5 = Assignment	5.1 = Assignable vehicle & person trip tables	TOD period, mode, OD pair
	5.2 = Route choice and network loading	TOD period, mode, OD pair
	5.3 = Level of service skimming	TOD period, mode, OD pair

Decision-making paradigm underlying AgBM can be summarized as follows:

- Individuals generate and preliminary schedule activities within the day segments.
- Tours are formed for each individual within day segments and cannot overlap in time. Trip destination, TOD, and mode choices are coordinated within the tour.
- All tours and trips are feasible in time and space.
- For each day segment the underlying sequence of choices is Activity allocation → Activity sequencing → Activity location → Tour formation.
- For each generated tour the underlying sequence of choices is Tour TOD → Tour mode combination → Within-tour time allocation and trip departure time → Car allocation and use.
- Segmentation is unlimited like in the case of ABM.
- Decision-making subjects can change parameters and adapt to new conditions.

3.6   Accounting for Time-Space Constraints and Model System Consistency

Different travel dimensions are intertwined and affect each other. Each new modeling paradigm tried to address a set of new aspects associated with model system integrity and internal consistency as summarized in Table 6.

Table 6. Internal consistency of travel demand.

Aspect	4-step	Tour-based	ABM	AgBM
Zonal trip ends match production & attraction constraints for each segment	Yes	Yes	Yes	Yes
Zonal production and attraction constraints for Home-based and Non-home-based trips match each other	No	Yes	Yes	Yes
Trip modes are consistent across trip chains	N/A	Yes	Yes	Yes

(*continued*)

Table 6. (*continued*)

Aspect	4-step	Tour-based	ABM	AgBM
Activities and tours for the same person do not overlap in time	N/A	N/A	Yes	Yes
Activities and tours for the same person are feasible in time & space	N/A	N/A	No	Yes
Car allocation and use within the household consistent with mode choice	N/A	N/A	N/A	Yes
Zonal car trip ends are consistent with parking capacity	N/A	N/A	Yes	Yes

3.7   Specifics of Travel Choices and Requirements for Statistical Models

Statistical models are applied at different steps of the decision-making process in all travel models. Historically, applied travel models were dominated by traditional econometric models of discrete choice based on the Random Utility Modeling (RUM) concept [1, 2]. Recently, a variety of Machine Learning (ML) methods were tried (k-nearest neighbors, random forest, neural networks). The pros and cons of RUM versus ML in the context of modeling travel behavior can be summarized as follows.

Traditional econometric models are characterized by the following features [1, 2]:

- Questionable fit at the individual level due to limitations of linear-in-parameters utility,
- Easy to calibrate to aggregate targets,
- Nice analytical properties (negative cross-elasticity) understood by practitioners,
- Well established utility specs and defaults for parameters like Value of Time (VOT) that correspond to expected model sensitivity to policies [13].

Machine Learning methods are characterized by the following features [14]:

- Better and in some cases, much better individual fit due to a consideration of combinations of variables with non-linear effects,
- More difficult to calibrate to aggregate targets,
- Analytical properties and elasticities are more complex and less understood by practitioners.
- No established culture of model specs or parameter values yet and some surprises with model sensitivity were reported [15].

Further analysis and comparison of RUM and ML methods is recognized as an important strategic avenue in modeling travel behavior especially within the AgBM paradigm [14, 15].

4   Conclusions

In general, there are two main directions for a further improvement of applied travel models that reflect two types of "truth" that can be observed. The first direction can be called "behavioral realism" and it relates to the best possible reflection on underlying individual decision making process and observed individual behavior. The second direction may look completely different and includes a better aggregate validation of the model with respect to traffic counts, transit ridership, and observed "big data" patterns of trip distribution. The first direction is preferred in academic research on travel behavior while the second direction is largely preferred by practitioners. At this moment. the grand unification theory does not exist yet. However, several important avenues can be outlined:

- Analysis and understanding of individual travel behavior remains the cornerstone of travel demand models. Such aspects as the necessary rich segmentation of individuals and activities as well as the necessity to address time-space constraints on travel that are largely ignored in aggregate trip-based and tour-based models seem to be a part of consensus [1, 2, 6, 12].
- However, from the practical perspective, it is important to demonstrate the aggregate validity of the travel model in terms of replication of the observed aggregate data such as traffic counts, transit ridership, and Origin-Destination trips patterns available today from the "big data" providers. This reconciliation of the highly disaggregate behavioral model with the aggregate controls is considered as the main stumbling block on the way of the advanced ABMs and AgBMs into practice with a complete replacement of the aggregate trip-based and tour-based models [1, 3]. Effective using of aggregate proportions as a basis for modeling (drawing) individual choices (so-called "data-driven" approach) is especially appealing for spatial trip distribution where advantage of "big data" can be taken [2].
- A practical aspect that humpers a wide-spread of advanced ABMs and AgBMs is long runtime and complexity of the software for the public agencies to handle [2, 12]. Long runtime issue is exacerbated by the necessity to equilibrate the travel model.
- Travel behavior and associated decision-making processes are inherently individual. However, there are important reasons why a travel model still needs to address aggregate macro constraints and factors that calls to an iterative equilibration structure. The main reason is an essential demand-supply equilibration due to aggregate capacity constraints. There are two major aggregate capacity constraints imposed on individual travel behavior. The first one relates to a constrained activity supply that is expressed in a competition for jobs and student enrollment as well as in time-varying accessibility measures (opening and closing hours of establishments) that affect discretionary trips [6]. The second one relates to a constrained transportation network capacity that results in highway network congestion effects, transit network crowding effects, and parking constraints [7]. Additionally, learning from others and "contagion" effects, for example, with respect to individual mobility attributes and modality preferences [8] can also be effectively modeled through equilibration.

- Emerging disruptive technologies such as Autonomous Vehicles (AVs) can change travel behavior substantially and first attempts to incorporate AVs in travel models are underway. It was recognized that simplified aggregate trip-based and tour-based models as well as even standard ABMs are not flexible enough to portray such a principal change in mobility. Only AgBMs seem can provide a reasonable platform for modeling AVs [1, 12]. The core advantage of AgBM in this regard is that it allows for a meaningful separation of the activity generation and travel formation steps in decision-making where AVs mostly affect the travel arrangement while the fundamental activity participation needs will remain similar to the observed travel behavior [16].

References

1. Boyce, D., Williams, H.: Forecasting Urban Travel: Past, Present, and Future. Edward Elgar, Cheltenham (2015)
2. NCHRP Synthesis 406: Advanced Practices in Travel Forecasting. Transportation Research Board (2010)
3. Ferdous, N., Bhat, C., Vana, L., Schmitt, D., Bowman, J., Bradley, M., Pendyala, R.: Comparison of Four-Step versus Tour-Based Models in Predicting Travel Behavior before and after Transportation System Changes – Results Interpretation and Recommendations. FHWA (2011)
4. Ye, X., Pendyala, R.M., Gottardi, G.: An exploration of the relationship between mode choice and complexity of trip chaining patterns. Transp. Res. Part B **41**(1), 96–113 (2007)
5. Vovsha, P.: Microsimulation travel demand models in practice in the US and prospects for agent based approach. In: Highlights of Practical Applications of Cyber-Physical Multi-Agent Systems. Proceedings of the International Workshops of PAAMS 2017, Porto, Portugal, 21–23 June, pp. 52–68. Springer (2017)
6. Bhat, C.R., Goulias, K.G., Pendyala, R.M., Paleti, R., Sidharthan, R., Schmitt, L., Hu, H.-H.: A household-level activity pattern generation model with an application for Southern California. Transportation **40**(5), 1063–1086 (2013)
7. Vovsha, P., Hicks, J.E., Anderson, R., Giaimo, G., Rousseau, G.: Integrated model of travel demand and network simulation. In: Proceedings of the 6th Conference on Innovations in Travel Modeling (ITM), TRB, Denver, CO (2016)
8. Dugundji, E., Walker, J.: Discrete choice with social and spatial network interdependencies: an empirical example using mixed generalized extreme value models with field and panel effects. Transp. Res. Rec. **1921**, 70–78 (2005)
9. Lemp, J.: Understanding joint daily activity pattern choices across household members using a latent class model framework. In: 93rd Annual TRB meeting (2014)
10. Vuk, G., Bowman, J., Daly, A.J., Hess, S.: Impact of family in-home quality time on person travel demand. Transportation **43**(4), 705–724 (2016)
11. Vovsha, P., Gliebe, J., Petersen, E., Koppelman, F.: Comparative analysis of sequential and simultaneous choice structures for modeling intra-household interactions. In: Timmermans, H. (ed.) Progress in Activity-Based Analysis, pp. 223–258. Elsevier Science Ltd, Oxford (2005)

12. Zhang, L., Chang, G.-L., Asce, M., Zhu, S., Xiong, C., Du, L., Mollanejad, M., Hopper, N., Mahapatra, S.: Integrating an agent-based travel behavior model with large-scale microscopic traffic simulation for corridor-level and subarea transportation operations and planning applications. Urban Plann. Dev. **139**, 94–103 (2013)
13. Paleti, R., Vovsha, P., Givon, D., Birotker, Y.: Impact of individual daily travel pattern on value of time. Transportation **42**(6), 1003–1017 (2015)
14. Hagenauer, J., Helbich, M.: A Comparative study of machine learning classifiers for modeling travel mode choice. Expert Syst. Appl. **78**, 273–282 (2017)
15. Golshani, N., Shabanopour, R., Mahmoudifard, S.M., Derrible, S., Mohammadian, A.: Comparison of artificial neural networks and statistical copula-based joint models. Presented at the 96th Annual Meeting of the Transportation Research Board, Washington, DC. (2017)
16. Maciejewsky, M., Bischoff, J., Horl, S., Nagel, K.: Towards a testbed for dynamic vehicle routing algorithms. In: Highlights of Practical Applications of Cyber-Physical Multi-Agent Systems. Proceedings of the International Workshops of PAAMS 2017, Porto, Portugal, 21–23 June 2017, pp. 69–79. Springer (2017)

Formalisation of Situated Dependent-Type Theory with Underspecified Assessments

Roussanka Loukanova[(✉)]

Stockholm University, Stockholm, Sweden
rloukanova@gmail.com

Abstract. We introduce a formal language of situated dependent-type theory, by extending its potentials for structured data that is integrated with quantitative assessments. The language has terms for situated information, which is partial and underspecified. The enriched formal language provides integration of a situated dependent-type theory with statistical and other approaches to machine learning techniques.

Keywords: Formal language · Information · Situations · Partiality
Memory variables · Information polarity · Underspecification

1 Backgrounds

The central ideas of Situation Theory was originally introduced by Barwise [1] and Barwise and Perry [3] Barwise [2], as a new approach to model theory of general information, which is relational, partial, and dependent on situations. An informal introduction to Situation Theory and Situation Semantics is presented by Devlin [5]. Rigorous introduction to mathematics of situated model theory is introduced by Seligman and Moss [13] and Loukanova [6]. Implementation work with logic programming for Situation Theory was restricted to first-order versions, e.g., see Tin and Akman [14,15].

Another, closely related approach to major concepts, on which computational models of information depend, is by a new theory of the mathematical notion of algorithm, originally introduced by Moschovakis [11]. Moschovakis [12] introduced its typed version, by a functional, formal language of recursion, which also represents context dependency of functional computations, by the notion of state dependent computations and function values. That line of work has been extended in various directions, e.g., see Loukanova [8,10].

This research has been supported, by covering my participation in DCAI 2018, by the Ministerio de Economía y Competitividad and the Fondo Europeo de Desarrollo Regional under the project number FFI2015-69978-P (MINECO/FEDER, UE) of the Programa Estatal de Fomento de la Investigación Científica y Técnica de Excelencia, Subprograma Estatal de Generación de Conocimiento.

E. Bucciarelli et al. (Eds.): DCAI 2018, AISC 805, pp. 49–56, 2019.
https://doi.org/10.1007/978-3-319-99698-1_6

Our paper is on a new, largely open topic of formalisation of Situation Theory, with computational syntax and semantic models of finely-grained information, initiated in Loukanova [7]. Here, we extend the formal language introduced in [7] in an essential way, by allowing parametric and numerical polarities, which can take values that are not only 0 and 1, but within a full range of rational or real numbers, e.g., in the interval [0, 1]. This allows polarities to be assessment values, by using methods of mathematical statistics. This approach can be developed for the extended Loukanova [9], but it requires more sophisticated methods, e.g., from machine learning, for the multi-dimensional neural structures.

Both higher-order, typed Situation Theory of information and type-theoretic formal languages for its versions are opening new areas of research. Computational semantics and computational neuroscience of language are among the primary applications of Situation Theory and classes of formal languages for it. Here, we introduce a higher-order, typed formal language of Situation Theory with memory variables that are subject to situation-theoretic restrictions and assignments. In addition, the language allows for integration with statistical and other approaches to development of new machine learning techniques that produce structured informational content. The structural part of language L_{ra}^{ST} introduced in this paper, is similar to that in Loukanova [9], by representing structured information about relations, properties, functional procedures, about objects standing in relations in situations, being subject to restrictions, and algorithmic computations. The restricted recursion terms can restrict what information can be saved in memory locations p individually, by constraints of the form $(p : T)$. The language L_{ra}^{ST} covers restrictions over sets of objects, e.g., by propositional constraints $(\{p_1, \ldots, p_m\} : T)$, where T is a term for a type with m argument roles. By such a generalized constraint, the memory variables $p_1, \ldots, p_m$ represent memory networks that are restricted to be simultaneously of the type T, for memorising such objects. Memory variables can be instantiated with more specialised information via specialized recursion terms. A generalised constraint over a set of instantiations requires that memory variables are simultaneously of a given type.

The contribution in this paper is the new treatment of the polarity type, its corresponding objects, which are numerical assessments. In our previous work on type-theoretic, situated-information theory, POL is the type for dual polarity, e.g., typically presented by the natural numbers 1 and 0, which express that 'the relation under the consideration holds or does not hold, respectively'. Note that the polarity is not a truth value. In this paper, we develop the notion of *relation polarity* to express more subtle information about objects standing or not in a given relation to some degree, and how to consolidate different assessment polarities over same objects in a given situation.

We generalise the concept of underspecified information in various aspects. Firstly, terms use specialised recursion variables, which are generalised, restricted memory variables for saving partially available information. Secondly, situations may have no information, or only partially available information. We generalise the notion of a relational polarity to be a numerical assessment of a degree to which a relation may hold for given objects. In addition, the degree itself can be undecided and parametrically available. The target of the work is mutual enrichment of techniques for processing partial and underspecified data.

2 Syntax of L_{ra}^{st}

2.1 Basic Types and Vocabulary of L_{ra}^{st}

The syntax of the extended formal language L_{ra}^{ST} is similar to that introduced in Loukanova [7]. Here, we extend it in an essential way, by allowing parametric and numerical polarities, which can take as values rational or real numbers, e.g., in the interval $[0, 1]$. This allows assessment values, i.e., polarities, by using methods of mathematical statistics.

The Types of L_{ra}^{ST} are defined recursively, starting with *primitive (basic) types*, BTypes = {IND, LOC, REL, FUN, POL, ARGR, INFON, SIT, PROP, PAR, TYPE, $\models$}.

IND is the type for individuals; LOC, for space-time locations; REL, for relations; FUN, for functions; TYPE, for types; PAR, for parameters; ARGR, for argument roles (slots); POL, for relation or property polarities; INFON, for objects that are information units; PROP, for objects that are propositions; SIT, for situations. The type $\models$ is designated and called "supports".

Constants: For each $\tau \in$ Types, L_{ra}^{ST} has a denumerable set of constants K_τ. The new feature of L_{ra}^{ST} is that we take constants K_{POL} semantically designating *polarity values* in a set K_{POL}, which can alternatively be, depending on applications of L_{ra}^{ST}, either rational $\mathbb{Q}$ or, as in (1), real numbers $\mathbb{R}$ in the closed interval $[0, 1]$:

$$K_{POL} \overset{1\text{-}1}{\rightarrowtail} K_{POL} \subset [0, 1] = \text{the set of the real numbers } i \in \mathbb{R},\ 0 \le i \le 1 \qquad (1)$$

While the set of the real numbers in $[0, 1]$ is uncountable, one can work with a countable set of constants K_{POL} covering possible polarity values in K_{POL}. In a semantic domain of a formal language, it is not necessary for every possible semantic object to be designated by a corresponding constant; $K = \bigcup_{\tau \in \text{Types}} K_\tau$.

Pure variables: For each type $\tau \in$ Types, L_{ra}^{ST} has a countable set of *pure variables*: PureVars$_\tau = \{ v_0^\tau, v_1^\tau, \dots \}$. They are used for binding by λ-abstraction.

Restricted recursion variables: For each type $\tau \in$ Types, L_{ra}^{ST} has a set of *recursion variables* or *(restricted) memory variables*: RecVars$_\tau = \{ p_0^\tau, p_1^\tau, \dots \}$ of the corresponding type. In particular, for each of the basic types, we take a set of restricted memory variables, which can be used for saving information and objects of the associated type.

As in Situation Theory introduced in Loukanova [6], the basic relations and types are associated with argument roles that have to satisfy constraints for their appropriate filling. We represent such constraints by using types.

Definition 1 (Argument roles with appropriateness constraints). *The formal language* L_{ra}^{ST} *respects situated semantic domains, e.g., in Loukanova [6].*

1. *Every relation constant, relation variable, and basic (i.e., primitive) type γ, $\gamma \in \mathcal{A}_{\mathrm{REL}} \cup \mathsf{Vars}_{\mathrm{REL}} \cup B_{\mathrm{TYPE}}$, is associated with a set $Args(\gamma)$, called the set of the argument roles of γ, so that:*

$$Args(\gamma) \equiv \{\, T_1 : \mathsf{arg}_1, \ldots, T_n : \mathsf{arg}_n \,\} \tag{2}$$

where $n \geq 0$, $\mathsf{arg}_1, \ldots, \mathsf{arg}_n \in \mathcal{A}_{\mathrm{ARGR}}$, and $T_1, \ldots, T_n$ are sets of types (basic or complex). The expressions (i.e., constants, or more complex expressions) $\mathsf{arg}_1, \ldots, \mathsf{arg}_n$, of type ARGR, are called the argument roles *(or the* names *of the argument slots) of γ. The sets of types $T_1, \ldots, T_n$ are specific for the argument roles of γ and are called the* basic appropriateness constraints *of the argument roles of γ.*

2. *Every function constant and variable γ, $\gamma \in \mathcal{A}_{\mathrm{FUN}} \cup \mathsf{Vars}_{\mathrm{FUN}}$, is associated with two sets: (1) $Args(\gamma)$, called the set of the argument roles of γ, and (2) $Value(\gamma)$, called the (singleton set of the) value role of γ, so that:*

$$Args(\gamma) \equiv \{\, T_1 : \mathsf{arg}_1, \ldots, T_n : \mathsf{arg}_n \,\} \tag{3a}$$
$$Value(\gamma) \equiv \{\, T_{n+1} : \mathsf{arg}_{n+1} \,\} \tag{3b}$$

where $n \geq 0$, $\mathsf{arg}_1, \ldots, \mathsf{arg}_{n+1} \in \mathcal{A}_{\mathrm{ARGR}}$, and $T_1, \ldots, T_{n+1}$ are sets of types (basic or complex). The expressions (i.e., constants, or more complex expressions), $\mathsf{arg}_1, \ldots, \mathsf{arg}_n$, of type ARGR, are called the argument roles *of γ. The expression arg_{n+1} is called the* value role *of γ. The sets of types $T_1, \ldots, T_{n+1}$ are specific for the argument roles of γ and are called, respectively, the* basic appropriateness constraints *of the arguments* and of the *value of γ.*

The set of all $\mathrm{L}_{\mathrm{ra}}^{\mathrm{ST}}$-terms is $\mathsf{Terms}(K) = \bigcup_{\tau \in \mathsf{Types}} \mathsf{Terms}_\tau$, where the sets Terms_τ are defined recursively as follows.

2.2 The Terms of $\mathrm{L}_{\mathrm{ra}}^{\mathrm{st}}$

Constants. If $c \in K_\tau$, then $c \in \mathsf{Terms}_\tau$, (i.e., every constant of type τ is also a term of type τ) denoted as $c : \tau$; $\mathsf{FreeV}(c) = \varnothing$ and $\mathsf{BoundV}(c) = \varnothing$.

Variables. For each type $\tau \in \mathsf{Types}$, $\mathsf{Vars}_\tau = \mathsf{PureVars}_\tau \cup \mathsf{RecVars}_\tau$. If $x \in \mathsf{Vars}_\tau$, then, $x \in \mathsf{Terms}_\tau$, denoted $x : \tau$; $\mathsf{FreeV}(x) = \{\, x \,\}$ and $\mathsf{BoundV}(x) = \varnothing$.

Infon Terms. For every relation term (basic or complex) $\rho \in \mathsf{Terms}_{\mathrm{REL}}$, associated with argument roles $Args(\rho) = \{\, T_1 : \mathsf{arg}_1, \ldots, T_n : \mathsf{arg}_n \,\}$, and terms ξ_1, $\ldots, \xi_n$ such that $\xi_1 : T_1, \ldots, \xi_n : T_n$, for every space-time location term $\tau : \mathrm{LOC}$, and every polarity term $t : \mathrm{POL}$, the expression in (4) is an *infon term*:

$$\begin{aligned} \ll \rho, T_1 : \mathsf{arg}_1 : \xi_1, \ldots, T_n &: \mathsf{arg}_n : \xi_n, \\ \mathrm{LOC} : Loc : \tau, \; \mathrm{POL} &: Pol : t \gg : \mathrm{INFON} \end{aligned} \tag{4}$$

All free (bound) occurrences of variables in $\rho, \xi_1, \ldots, \xi_n, \tau$ are also free (bound) in the infon term. By agreement, we often use alternative, understandable notations of the infon terms, assuming that there is no confusion.

Proposition Terms. For every type term (basic or complex) $\gamma \in \mathsf{Terms}_{\mathrm{TYPE}}$, associated with argument roles $Args(\gamma) \equiv \{\, T_1 : \mathsf{arg}_1, \ldots, T_n : \mathsf{arg}_n \,\}$, and terms $\xi_1, \ldots, \xi_n$ that satisfy the corresponding appropriateness constraints of the argument roles of γ, $\xi_1 : T_1, \ldots, \xi_n : T_n$, the expression in (5) is a *full proposition term*, including its type association to PROP. It is used when the type labeling is relevant, and can be skipped if not needed.

$$(\gamma, T_1 : \mathsf{arg}_1 : \xi_1, \ldots, T_n : \mathsf{arg}_n : \xi_n, \mathrm{POL} : Pol : i) : \mathrm{PROP} \tag{5}$$

All free (bound) occurrences of variables in γ, $\xi_1, \ldots, \xi_n$, τ are also free (bound) in the proposition term.

Often, depending on the context, we use appropriate notations and abbreviations.

Application Terms. For every function term, basic or complex, $\gamma \in \mathsf{Terms}_{\mathrm{FUN}}$, that is associated with argument roles $Args(\gamma) \equiv \{\, T_1 : \mathsf{arg}_1, \ldots, T_n : \mathsf{arg}_n \,\}$ and with a value role $Value(\gamma) \equiv \{\, T_{n+1} : \mathsf{val} \,\}$, and for every terms $\xi_1, \ldots, \xi_n, \xi_{n+1}$ that satisfy the corresponding appropriateness constraints of the argument and value roles of γ, the expression in (6), is an *application term*:

$$\gamma\{T_1 : \mathsf{arg}_1 : \xi_1, \ldots, T_n : \mathsf{arg}_n : \xi_n\} : T_{n+1} \tag{6}$$

All free (bound) occurrences of variables in γ, $\xi_1, \ldots, \xi_n$, τ are also free (bound) in the application terms.

λ-abstraction Terms Case 1: complex relations with complex argument roles. For every infon term $I : \mathrm{INFON}$ (basic or complex) and any pure variables $\xi_1, \ldots, \xi_n \in \mathsf{PureVars}$ (which may occur freely in I), the expression $\lambda\{\xi_1, \ldots, \xi_n\}\, I$ is a *complex-relation term*, i.e.:

$$\lambda\{\xi_1, \ldots, \xi_n\}\, I : \mathrm{REL} \tag{7}$$

The argument roles of $\lambda\{\xi_1, \ldots, \xi_n\}I$, which we denote by $[\xi_1], \ldots, [\xi_n]$, are associated with corresponding *appropriateness constraints*, as follows:

$$Args(\lambda\{\xi_1, \ldots, \xi_n\}\, I) \equiv \{\, T_1 : [\xi_1], \ldots, T_n : [\xi_n] \,\} \tag{8}$$

where, for each $i \in \{1, \ldots, n\}$, T_i is the union of all types that are the appropriateness constraints of all the argument roles occurring in I, such that ξ_i fills up them, without being bound. (Note that ξ_i may fill more than one argument role in I.)

Case 2: complex types with complex argument roles. For every proposition term $\theta : \mathrm{PROP}$ (basic or complex) and any pure variables $\xi_1, \ldots, \xi_n \in \mathsf{PureVars}$ (which may occur freely in θ, in the interesting cases), the expression $\lambda\{\xi_1, \ldots, \xi_n\}\, \theta$ is a *complex-type term*, i.e.:

$$\lambda\{\xi_1, \ldots, \xi_n\}\, \theta : \mathrm{TYPE} \tag{9}$$

The argument roles of $\lambda\{\xi_1, \ldots, \xi_n\}\theta$, which we denote by $[\xi_1], \ldots, [\xi_n]$, are associated with corresponding *appropriateness constraints* that are the types $T_1, \ldots, T_n$, respectively, i.e., $\{\, T_1 : [\xi_1], \ldots, T_n : [\xi_n] \,\}$, as follows:

$$Args(\lambda\{\xi_1, \ldots, \xi_n\}\, \theta) \equiv \{\, T_1 : [\xi_1], \ldots, T_n : [\xi_n] \,\} \tag{10}$$

where, for each $i \in \{1, \ldots, n\}$, T_i is the union of all types that are the appropriateness constraints of all the argument roles occurring in θ, such that ξ_i fills up them, without being bound. (Note that ξ_i may fill more than one argument role in θ.)

Case 3: complex function terms (operation terms) with complex arguments. For every term $\varphi \in \mathsf{Terms}_\tau$ (basic or complex), where $\tau \in \mathsf{Types}$, $\tau \not\equiv \text{INFON}$, $\tau \not\equiv \text{PROP}$, and for all pure variables $\xi_1, \ldots, \xi_n \in \mathsf{PureVars}$ (which may occur freely in φ), the expression $\lambda\{\xi_1, \ldots, \xi_n\}\varphi$ is a *complex-function term*, i.e.:

$$\lambda\{\xi_1, \ldots, \xi_n\}\varphi : \text{FUN} \tag{11}$$

The term $\lambda\{\xi_1, \ldots, \xi_n\}\varphi$ has a value role, $Value(\varphi) \equiv \{\tau : \mathsf{val}\}$, and argument roles, denoted by $[\xi_1], \ldots, [\xi_n]$, which are associated with corresponding *appropriateness constraints* as follows:

$$Args(\lambda\{\xi_1, \ldots, \xi_n\}\varphi) \equiv \{T_1 : [\xi_1], \ldots, T_n : [\xi_n]\} \tag{12}$$

where, for each $i \in \{1, \ldots, n\}$, T_i is the union of all types that are the appropriateness constraints of all the argument roles occurring in φ, such that ξ_i fills up them, without being bound. (Note that ξ_i may fill more than one argument role in φ.)

In the above Cases 1, 2, 3, all the free occurrences of $\xi_1, \ldots, \xi_n$ in φ are bound in the term $\lambda\{\xi_1, \ldots, \xi_n\}\phi$. All other free (bound) occurrences of variables in φ are free (bound) in the term $\lambda\{\xi_1, \ldots, \xi_n\}\varphi$.

Restricted Recursion Terms For any $\mathsf{L}^{\text{ST}}_{\text{ra}}$-terms for types $\mathsf{C}_k : \text{TYPE}$, with l_k argument roles, i.e., $Args(\mathsf{C}_k) \equiv \{T_{k,1} : arg_{k,l_1}, \ldots, T_{k,l_k} : arg_{k,l_k}\}$ and pairwise different memory variables $\overrightarrow{\mathsf{q}_k} \in \mathsf{RecVars}_{T_k}$, for $k = 1, \ldots, m$, and for any terms $\mathsf{A}_i : \sigma_i$, for $i = 0, \ldots, n$, and pairwise different memory variables $\mathsf{p}_i \in \mathsf{RecVars}_{\sigma_i}$, for $i = 1, \ldots, n$, the expression (13) is a *restricted recursion term* of type σ_0:

$$\begin{aligned} &[\mathsf{A}_0 \text{ such that } \{(\mathsf{q}_{1,1}, \ldots, \mathsf{q}_{1,l_1} : \mathsf{C}_1), \ldots, (\mathsf{q}_{m,1}, \ldots, \mathsf{q}_{m,l_m} : \mathsf{C}_m)\}] \\ &\text{where } \{\mathsf{p}_1 := \mathsf{A}_1, \ldots, \mathsf{p}_n := \mathsf{A}_n\} \in \mathsf{Terms}_{\sigma_0} \end{aligned} \tag{13}$$

All free occurrences of $\mathsf{p}_1, \ldots, \mathsf{p}_n$ in $\mathsf{A}_0, \ldots, \mathsf{A}_n, \mathsf{C}_1, \ldots, \mathsf{C}_m$, are bound in the term (13). All other free (bound) occurrences of variables are free (bound) in (13).

The expression (13) is *acyclic recursion term*, if the sequence of assignments $\{\mathsf{p}_1 := \mathsf{A}_1, \ldots, \mathsf{p}_n := \mathsf{A}_n\}$ satisfies the acyclicity constraint given in Definition 2.

Definition 2 (Acyclicity Constraint). *A sequence of assignments*

$$\{\mathsf{p}_1 := \mathsf{A}_1, \ldots, \mathsf{p}_n := \mathsf{A}_n\}$$

is acyclic iff there is a ranking function $\mathsf{rank} : \{\mathsf{p}_1, \ldots, \mathsf{p}_n\} \longrightarrow \mathbb{N}$ *such that, for all* $\mathsf{p}_i, \mathsf{p}_j \in \{\mathsf{p}_1, \ldots, \mathsf{p}_n\}$, *if* p_j *occurs freely in* A_i, *then* $\mathsf{rank}(\mathsf{p}_j) < \mathsf{rank}(\mathsf{p}_i)$. *Note that the acyclicity constraint is a proper part of the recursive definition of the* $\mathsf{L}^{\text{ST}}_{\text{ra}}$-*terms.*

2.3 Propositional Content with Uncertainty

While the term A, in (14b)–(14e), is simple, for expository purpose, it demonstrates the technical notions introduced in the first part of the paper. The term A is a pattern for situated information. The specialised recursion variables, i.e., memory variables, x, b, y, l, and p would have been fully underspecified without the components (14c)–(14e) of the term A. The component (14c) provides constraints, while the assignments in (14d)–(14e) specify algorithmic computations (very simple in this example). The head part (14a)–(14b) of the term A is a proposition, which states that an individual x invests amount b, in an undertaking y, at a location l. The restrictor component (14c) states that the location l of the investment is sub-location of the location l_1, in which the amount b has realisation. The recursion assignments in (14d)–(14e) specify the variables correspondingly, by recursive computations (very simple in this example). Our focus is on the assignment in (14e), which expresses that the investment information in the situation s is available with uncertainty, e.g., probabilistic measure, 0.75, i.e., 75%. If the assignment has been $p := 0$, the situation s carries information that the investment is not realised at the location l; with $p := 1$, that it is realised.

$$A \equiv (s \models \ll \textit{invest, inverster} : x, \textit{invested} : b, \textit{undertaking} : y, \qquad (14a)$$

$$\textit{Loc} : l; p \gg \qquad\qquad\qquad$$

$$\wedge \ll \textit{amount, arg} : b, \textit{Loc} : l_1; p \gg) \qquad (14b)$$

$$\textsf{such that } \{\ (l \subset l_1)\} \qquad\qquad (14c)$$

$$\textsf{where } \{\ x := \textit{John}, \ b := (b_1 + b_2), \ b_1 := 1000\,\text{€}, \ b_2 := 5000\,\text{€}, \qquad (14d)$$

$$p := 0.75\ \} \qquad\qquad\qquad (14e)$$

3 Future Work

We introduce dependent-type theory of situated information provided with quantitate assessments. Distinctively new features are that information is structured and enriched with situated content about objects and relations between them according to numerical values. Our development of the situated type-theory targets integration of statistical approaches to information and machine learning techniques with dependent type-theory of situated information. The theory provides structured, situated information with parameters that can be fully or partly underspecified and dynamically updated. The updates are facilitated by the recursion terms with constraints and assignments. The next stages of our work include development of the notions of canonical forms and reduction calculus to canonical forms. Another line of work is finding suitable quantitate approach from mathematical statistics and machine learning for the integration, depending on areas of applications.

References

1. Barwise, J.: Scenes and other situations. J. Philos. **78**, 369–397 (1981)
2. Barwise, J.: The Situation in Logic. No. 17 in CSLI Lecture Notes. CSLI Publications, Stanford, California (1989)
3. Barwise, J., Perry, J.: Situations and Attitudes. MIT Press, Cambridge (1983). republished as [4]
4. Barwise, J., Perry, J.: Situations and Attitudes, The Hume Series. CSLI Publications, Stanford, California (1999)
5. Devlin, K.: Situation theory and situation semantics. In: Gabbay, D., Woods, J. (eds.) Handbook of the History of Logic, vol. 7, pp. 601–664. Elsevier (2008). http://web.stanford.edu/~kdevlin/Papers/HHL_SituationTheory.pdf
6. Loukanova, R.: Situation theory, situated information, and situated agents. In: Nguyen, N.T., Kowalczyk, R., Fred, A., Joaquim, F. (eds.) Transactions on Computational Collective Intelligence XVII, Lecture Notes in Computer Science, vol. 8790, pp. 145–170. Springer, Heidelberg (2014). https://doi.org/10.1007/978-3-662-44994-3_8
7. Loukanova, R.: Underspecified relations with a formal language of situation theory. In: Loiseau, S., Filipe, J., Duval, B., van den Herik, J. (eds.) Proceedings of the 7th International Conference on Agents and Artificial Intelligence, vol. 1, pp. 298–309. SciTePress — Science and Technology Publications, Lda (2015). https://doi.org/10.5220/0005353402980309
8. Loukanova, R.: Acyclic recursion with polymorphic types and underspecification. In: van den Herik, J., Filipe, J. (eds.) Proceedings of the 8th International Conference on Agents and Artificial Intelligence, vol. 2, pp. 392–399. SciTePress — Science and Technology Publications, Lda (2016). https://doi.org/10.5220/0005749003920399
9. Loukanova, R.: Typed theory of situated information and its application to syntax-semantics of human language. In: Christiansen, H., Jiménez-López, M.D., Loukanova, R., Moss, L.S. (eds.) Partiality and Underspecification in Information, Languages, and Knowledge, pp. 151–188. Cambridge Scholars Publishing (2017)
10. Loukanova, R.: Gamma-star reduction in the type-theory of acyclic algorithms. In: Rocha, A.P., van den Herik, J. (eds.) Proceedings of the 10th International Conference on Agents and Artificial Intelligence (ICAART 2018), vol. 2, pp. 231–242. INSTICC, SciTePress — Science and Technology Publications, Lda (2018). https://doi.org/10.5220/0006662802310242
11. Moschovakis, Y.N.: The logic of functional recursion. In: Logic and Scientific Methods, pp. 179–207. Kluwer Academic Publishers/Springer (1997)
12. Moschovakis, Y.N.: A logical calculus of meaning and synonymy. Linguist. Philos. **29**(1), 27–89 (2006). https://doi.org/10.1007/s10988-005-6920-7
13. Seligman, J., Moss, L.S.: Situation theory. In: van Benthem, J., ter Meulen, A. (eds.) Handbook of Logic and Language, pp. 253–329. Elsevier, Amsterdam (2011)
14. Tin, E., Akman, V.: Information-oriented computation with BABY-SIT. In: Seligman, J., Westerståhl, D. (eds.) Logic, Language and Computation, vol. 1, pp. 19–34. No. 58 in CSLI Lecture Notes. CSLI Publications, Stanford (1996)
15. Tın, E., Akman, V.: Situated nonmonotonic temporal reasoning with baby-sit. AI Commun. **10**(2), 93–109 (1997)

Scholarship, Admission and Application of a Postgraduate Program

Yehui Lao[1,2(✉)], Zhiqiang Dong[1], and Xinyuan Yang[2]

[1] Key Lab for Behavioral Economic Science & Technology, South China Normal University, Guangzhou 510631, Guangdong, China
yehui.lao@foxmail.com,dongzhiqiang@m.scnu.edu.cn
[2] Department of Economics, University of Waterloo, 200 University Avenue West, Waterloo, ON N2L 3G1, Canada

Abstract. This paper aims to construct a game of admission and application behavior of a postgraduate program. It attempts to expand the decision of graduate school from one party model to two party model. It suggests that the interval of postgraduate program's scholarship determines the decision made by applicants with different capacity and family background. Also, graduate schools will try to use scholarship as a tool to select students.

Keywords: Decision making · Scholarship level
Postgraduate program

1 Introduction

Recently, a volume of literatures had revealed the choice of entering a postgraduate program and find that deciding to attend a postgraduate program is a risky strategy (Golde 2005; Powell and Green 2007). To explain why people choose this risky strategy, Brailsford (2010) investigated in Australia and found that a postgraduate degree may improve intrinsic interest of candidates. This finding validates some categories identified in the limited previous literature (Churchill and Sanders 2007; Gill and Hoppe 2009). Lindley and Machin (2016) found that postgraduate degree holders earned around 42–45% more than college-only graduates employee in routine intensive jobs in 2012. However, there is a lack of literature focusing on the admission behaviour of graduate school.

Agent-based modelling (ABM) is a computational approach to model and simulate agents' decision-making (Macal and North 2008). However, few literature analyze school choice with ABM. Henrickson (2002) designed an ABM to analyze the college choice/college access problem. Reardon et al. (2016), extending Hendrickson's work, simulated the college application and selection process with ABM. They found that relative could have significant impacts on students' choice of colleges.

The purpose of this paper is to expand the decision of graduate school, based on existing literatures, to a game of admission and application of a postgraduate

© Springer Nature Switzerland AG 2019
E. Bucciarelli et al. (Eds.): DCAI 2018, AISC 805, pp. 57–66, 2019.
https://doi.org/10.1007/978-3-319-99698-1_7

program. Two parties, including prospective students and graduate schools, are involved in this model. The main findings are that decisions made by applicants with different capacity and family background are determined by the interval of postgraduate program's scholarship. The scholarship is a useful tool of graduate schools to select students.

2   Model

2.1   Setting

It is a game model with incomplete information to the graduate school, while prospective applicants own the information. The time $t \in \{0, 1\}$. 0 represents the current time period which is enough to finish a graduate program while 1 is denoted as the future. There are two parties in this game: graduate schools (denoted as g) and prospective applicants (denoted as a). The wealth status of applicants' families (f) and capacity (c) are employed to measure the quality of applicant. The capacity is a binary variable containing high level (h) and low level (l)[1] while the wealth level of family is normalized. If the wealth level of family is rich, $f=0$, or $f=1$. Given $a \in \{{}^c_f\}$ and $c \in \{h, l\}$, $f \in \{1, 0\}$, thus there are four types of prospective applicant:$a \in \{\binom{h}{0}\binom{h}{1}\binom{l}{0}\binom{l}{1}\}$. The Nature determines the type of prospective applicant. Moreover, the capacity is highly related to the possibility of excellent academic attainment, which is one of the graduate degree requirements. To simplify the model, we assume that acquiring the excellent academic attainment is the only one requirement for the graduate degree. P_h and P_l are attaining attainment possibilities of prospective applicants with high capacity and with low capacity, given that $1 > P_h > P_l > 0$. The w_{it} is the wage level of an individual with(/without) graduate degree in the period t. If an individual attend a job currently without a graduate degree, she enjoys the current wage level w_{l0} and will have w_{l1} in the future. If, on the other hand, an individual enrolls in a graduate program, she will be likely to earn w_{h1} in the future. Also, she enjoys the scholarship sponsored by the graduate school in the period 0. However, she will earn w_{l1} if she fails the graduate program. Due to she choose to enroll in the program, she may feel the economic pressure in the period 0 if she is in a poor family. There is a set of applicant's strategies containing to attend a job and to apply for a postgraduate program.

On the other hand, the graduate school enjoys academic attainment of the student's because it will improve the rank and reputation of the school. In order to attract student with potential enroll in, the graduate school will sponsor students with scholarship, covering parts of living expenses.

---

[1] The capacity in this paper is related to the possibility of acquiring high quality publication. The reserved capacity is higher than the requirement of postgraduate program's admission.

2.2   Solution

The Payoff of a prospective applicant a can be written as follows:

$$U_{a1}^{cf} = w_{l0} + \frac{w_{l1}}{(1+\beta)} \tag{1}$$

$$U_{a2}^{cf} = (1 - P_c)\frac{w_{l1}}{(1+\beta)} + P_c\frac{w_{h1}}{(1+\beta)} - e \times f + S \tag{2}$$

The Eq. (1) is the payoff of attending to job currently without a graduate degree while the Eq. (2) is the expected payoff of enrolling in a graduate program. In the Eq. (2), $P_c\frac{w_{h1}}{(1+\beta)}$ is the expected wage level of graduate degree holders while $(1 - P_c)\frac{w_{l1}}{(1+\beta)}$ is the expected wage level of the scenario that applicants fail the postgraduate program. The f represents the wealth level of family and $f \in \{1, 0\}$. If the wealth level of family is rich, $f=0$, or $f=1$. The e denotes the economic pressure in the period 0. The term $e \times f$ implies that prospective applicants who live in a wealthy family are more able to withstand short-term economic pressures than people who live in a poor family. c represents the capacity of individual and $c \in \{h, l\}$. Finally, the β is the discount factor. The S is the scholarship level decided by two factor: the benchmark scholarship level and the percentage of scholarship.

Definition 1. The cost of taking a postgraduate program includes opportunity cost and the living expenses. The opportunity cost of taking a postgraduate program is the current wage for the position without graduate degree. Also, we can treat the living expenses as the economic pressure for the people who live in a poor family. So we can write it as: $D = w_{l0} + e$

Assumption 1 $w_{h1} \geq w_{l1} \geq w_{l0}$.

This Assumption ensures that the choice of entering a graduate program can be possible attributed to postgraduate degree holders earn the wage premium. In fact, the postgraduate wage premium related to college-only in United States has increase since 1979. (Lindley and Machin 2016) Furthermore, the w_{l1} is not less than w_{l0} because the wage level rise with the accumulation of experience. (Mincer 1958, 1974).

If the prospective applicant decides to attend to work currently, it must satisfy the condition that Eq. (1) > Eq. (2). That is: $w_{l0} + e \times f > P_c\frac{w_{h1}-w_{l1}}{1+\beta} + S$. It means the current wage level of job position without graduate degree is higher than the sum of the expected wage premium of a postgraduate degree and scholarship level. If the prospective applicant has the poor family, she also takes the economics pressure into consideration. Otherwise, the condition of attend to a graduate program is $w_{l0} + e \times f < P_c\frac{w_{h1}-w_{l1}}{1+\beta} + S$.

The payoff of graduate school is:

$$U_g(S|P_c) = P_c \times A - S \tag{3}$$

A is the utility of graduate school acquiring from students' academic attainment and $A \in (0, \infty)$. The Eq. (3) implies that the optimal rule of graduate school is to acquire high academic reputation and rank with low scholarship. We assume the scholarship is the only way for a graduate school attracting applicants.

Definition 2. The value of a postgraduate degree contains students' academic attainment and the wage level premium of a postgraduate degree. We can write it as follow: $V = A + \frac{w_{h1} - w_{l1}}{1 + \beta}$.

Lemma 1. Prospective applicants with both high capacity and wealthy families background have more incentive to enter a postgraduate program than others.

Proof of Lemma 1: When $c \in \{h, l\}$ and $P_h > P_l$, we have $U_{a2}^{hf} = (1 - P_h)\frac{w_{l1}}{(1 + \beta)} + P_h\frac{w_{h1}}{(1 + \beta)} - e \times f + S > (1 - P_l)\frac{w_{l1}}{(1 + \beta)} + P_l\frac{w_{h1}}{(1 + \beta)} - e \times f + S = U_{a2}^{lf}$. When $f \in \{1, 0\}$, we have $U_{a2}^{c0} = (1 - P_c)\frac{w_{l1}}{(1 + \beta)} + P_c\frac{w_{h1}}{(1 + \beta)} + S > (1 - P_c)\frac{w_{l1}}{(1 + \beta)} + P_c\frac{w_{h1}}{(1 + \beta)} - e + S = U_{a2}^{c1}$. As a result, U_{a2}^{h0} is the highest value among all type of U_{a2}^{cf}.

Lemma 2. If the economic pressure on students with poor families is higher than the gap of expected postgraduate wage premium between high capacity people and low capacity people, prospective applicants $a = \binom{l}{0}$ would have more interest in a postgraduate program than the prospective applicants $a = \binom{h}{1}$, otherwise they have less incentive.

Proof of Lemma 2: When $e > P_h\frac{w_{h1} - w_{l1}}{1 + \beta} - P_l\frac{w_{h1} - w_{l1}}{1 + \beta}$, we can rewrite it as $P_l\frac{w_{h1} - w_{l1}}{1 + \beta} + e > P_h\frac{w_{h1} - w_{l1}}{1 + \beta}$. When we add $\frac{w_{l1}}{(1 + \beta)} + S$ to both sides of equation, we have $(1 - P_l)\frac{w_{l1}}{(1 + \beta)} + P_l\frac{w_{h1}}{(1 + \beta)} + S > (1 - P_h)\frac{w_{l1}}{(1 + \beta)} + P_h\frac{w_{h1}}{(1 + \beta)} - e + S$, which also can be rewritten as $U_{a2}^{l0} > U_{a2}^{h1}$. So type $\binom{l}{0}$ have more incentive than type $\binom{h}{1}$.

Proposition 1. There is no applicant applying for the graduate program if and only if $S \in [0, w_{lo} - P_h\frac{w_{h1} - w_{l1}}{1 + \beta})$.

Proof of Proposition 1: Given the Lemma 1, we have $U_{a2}^{h0} \geq U_{a2}^{cf}, \forall c \in \{h, l\}$ and $f \in \{1, 0\}$. When $w_{lo} - P_h\frac{w_{h1} - w_{l1}}{1 + \beta} > S$, $U_{a1}^{cf} > U_{a2}^{h0}$, which means the strategy that enter a graduate program is the strictly dominated strategy to all types of applicant.

In general, even if the scholarship can not cover all opportunity cost of a graduate student, it should ensure the expected premium of a postgraduate degree is worthy.

Proposition 2. If there are a large number of prospective applicant for each type, the optimal interval scholarship is that $S \in [w_{lo} - P_h \frac{w_{h1} - w_{l1}}{1+\beta}, w_{lo} - P_l \frac{w_{h1} - w_{l1}}{1+\beta})$ when the economic pressure on students with poor families is not less than the gap of expected postgraduate wage premium between high capacity people and low capacity people.

Proof of Proposition 2: Given the Lemma 2, applicants $a = \binom{l}{0}$ have the second highest preference on entering a postgraduate program. According to the payoff function of the graduate school, only accepting type $\binom{h}{0}$ is the optimal decision when there are a large number of prospective applicant for each type. Therefore, graduate schools will give scholarship lower than a particular level to ensure only prospective applicants $\binom{h}{0}$ would apply for a postgraduate program. Type $\binom{l}{0}$ must choose to attend to job, which means $U_{a1}^{l0} > U_{a2}^{l0}$. We can expand this expression as $w_{lo} + \frac{w_{l1}}{(1+\beta)} > (1 - P_l)\frac{w_{l1}}{(1+\beta)} + P_l\frac{w_{h1}}{(1+\beta)} + S$. So $S < w_{lo} - P_l \frac{w_{h1} - w_{l1}}{1+\beta}$. According to the proof of proposition 1, $S > w_{lo} - P_h \frac{w_{h1} - w_{l1}}{1+\beta}$, we can have $S \in [w_{lo} - P_h \frac{w_{h1} - w_{l1}}{1+\beta}, w_{lo} - P_l \frac{w_{h1} - w_{l1}}{1+\beta})$.

The Proposition 2 implies that when $e \geq P_h \frac{w_{h1} - w_{l1}}{1+\beta} - P_l \frac{w_{h1} - w_{l1}}{1+\beta}$ and there are limited number of prospective applicant for each type, the preference of graduate school is: $U_g(\binom{h}{0}) \succeq U_g(\binom{l}{0}) \succeq U_g(\binom{h}{l}) \succeq U_g(\binom{l}{1})$. When there is no available type $\binom{h}{0}$ in the market, the optimal interval of scholarship will be expanded as $S \in [w_{lo} - P_h \frac{w_{h1} - w_{l1}}{1+\beta}, w_{lo} + e - P_l \frac{w_{h1} - w_{l1}}{1+\beta})$.

Based on the proof of Proposition 2, we can also have another result that the interval of scholarship is $S \in [w_{lo} - P_h \frac{w_{h1} - w_{l1}}{1+\beta}, w_{lo} - P_h \frac{w_{h1} - w_{l1}}{1+\beta} + e)$ when the economic pressure on students with poor families is less than the expected postgraduate wage premium gap between high capacity people and low capacity people. When $e < P_h \frac{w_{h1} - w_{l1}}{1+\beta} - P_l \frac{w_{h1} - w_{l1}}{1+\beta}$, the process of proof is similar to the proof of Proposition 2. We have $S \in (w_{lo} - P_h \frac{w_{h1} - w_{l1}}{1+\beta}, w_{lo} - P_h \frac{w_{h1} - w_{l1}}{1+\beta} + e]$ when there are a large number of prospective applicant for each type, while we have $S \in [w_{lo} - P_h \frac{w_{h1} - w_{l1}}{1+\beta} + e, w_{lo} - P_l \frac{w_{h1} - w_{l1}}{1+\beta})$ when there is no available type $\binom{h}{0}$ in the market.

To summarize, when there are a large number of prospective applicant for each type,we have:

$$s \in \begin{cases} [w_{lo} - P_h \frac{w_{h1}-w_{l1}}{1+\beta}, w_{lo} - P_l \frac{w_{h1}-w_{l1}}{1+\beta}), \ if \ e \geq P_h \frac{w_{h1}-w_{l1}}{1+\beta} - P_l \frac{w_{h1}-w_{l1}}{1+\beta} \\ [w_{lo} - P_h \frac{w_{h1}-w_{l1}}{1+\beta}, w_{lo} - P_h \frac{w_{h1}-w_{l1}}{1+\beta} + e), \ if \ e < P_h \frac{w_{h1}-w_{l1}}{1+\beta} - P_l \frac{w_{h1}-w_{l1}}{1+\beta} \end{cases}$$,

while when the type $\binom{h}{0}$ is not available in the market, we have:

$$s \in \begin{cases} [w_{lo} - P_l \frac{w_{h1}-w_{l1}}{1+\beta}, w_{lo} - P_h \frac{w_{h1}-w_{l1}}{1+\beta} + e), \ if \ e \geq P_h \frac{w_{h1}-w_{l1}}{1+\beta} - P_l \frac{w_{h1}-w_{l1}}{1+\beta} \\ [w_{lo} - P_h \frac{w_{h1}-w_{l1}}{1+\beta} + e, w_{lo} - P_l \frac{w_{h1}-w_{l1}}{1+\beta}), \ if \ e < P_h \frac{w_{h1}-w_{l1}}{1+\beta} - P_l \frac{w_{h1}-w_{l1}}{1+\beta} \end{cases}$$.

Proposition 3. If the possibility of student acquiring excellent academic attainments is less than the ratio of postgraduate program's cost to value of postgraduate degree, graduate schools will set up a maximum of scholarship to block students from poor family even they have high capacity.

Proof of Proposition 3: Given the Definition 1 and 2, $P_c < \dfrac{D}{V}$ can be rewritten it as $P_c < \dfrac{w_{lo}+e}{A+\frac{w_{h1}-w_{l1}}{1+\beta}}$. So

$$P_c \times A < w_{lo} - P_c \frac{w_{h1}-w_{l1}}{1+\beta} + e \tag{4}$$

Additionally, if $U_{a1}^{cl} < U_{a2}^{cl}$, people from poor family will apply for a postgraduate program. We have

$$S > w_{lo} - P_c \frac{w_{h1}-w_{l1}}{1+\beta} + e \tag{5}$$

Substitute Eq. (4) into Eq. (5), we have $S > P_c \times A$, which means $U_g < 0$, graduate schools will try to use Eq. (5) as the maximum scholarship ceiling to block students from poor family. In this proof, the capacity is not reflected.

3   Methodology

3.1   Design of the Simulation Model

We, based on the above theoretical models, simulate groups of participants, consisting four types of prospective applicants and the only one graduate school. The graduate school is consisted by members of committee, who share the same utility function and decide the percentage of benchmark scholarship.

The process is designed according to Moran process(as introduced in Nowak et al. 2004). At the beginning of each round, each participant is given a random state associated with her action. The initial probability of state is 50%. At the end of each round, each participant has a chance, based on the fitness, to "reproduce" (or"copy") an "offspring" with the same state in the next round, or be replaced by a newcomer, whose state is decided randomly based on the state distribution of the current round. The fitness function of prospective applicant is below:

$$Fitness_{ai} = \frac{U_{ai}}{\sum U_{ai}}, \forall i \in \{1, 2\} \qquad (6)$$

Participants are more likely to have "offspring" when they own a higher fitness. Furthermore, the fitness of graduate school's members is an exponential function of the payoff because there may be negative payoff. The fitness function of graduate school's member is below:

$$Fitness_{gj} = e^{wU_{gj}} \qquad (7)$$

In addition, there are binary actions for graduate school's members to decide whether the school provides scholarship or not. Then the percentage of scholarship is calculated and equals the percentage of approval of providing scholarship.

4 Results

The simulation was executed by Matlab, a mathematical software. A number of figures were produced by the progress of the simulation, including the four type of individuals' probability of applying for a postgraduate program over time, and the percentage of scholarship over time.

Thus far, we have discussed behaviour and payoffs of prospective applicant and graduate school, and the set of parameters which are required to complete the utility function for each player in the game. We used the following scenarios in the initial analysis, the results of which are summarized below:

We use two dimensions, as we mention before, to categorise prospective applicants based on Pareto principle. That means the distribution of rich-poor people is 20/80 while the high-low capacity people distribution is 20/80 as well(see Table 1). The number of graduate school's member is set as 100 for the reason of convenience. Members will vote, based on their own states, whether the graduate school provides scholarship or not. The shares of votes saying "yes" is the percentage of benchmark scholarship. The real scholarship level is that benchmark scholarship times the percentage, and the payoffs of applicant is based on the real scholarship level. This process runs 500 rounds. Furthermore, the current wage level is 100 while it will raise to 130 due to inflation. Lindley and Machin (2016) found that postgraduate degree holders, on average, earned around 23–24% more than college-only graduates employee in routine intensive jobs in 1980 and rose to around 42–45% in 2012. Therefore, we set the future wage level 162 in this subsection.

In the scenario of different types of applicants, the dynamics of prospective applicants' behaviour are captured by four different paths over time, which ultimately converge to different steady states as indicated in Fig. 1(a).

The graduate school is typically faced with different distributions from various types of applicants. To maximize the payoff, the graduate school provides

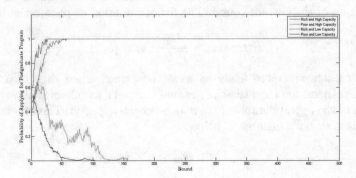

(a) The simulation of each type of applicant with different distribution

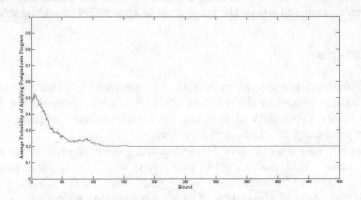

(b) The simulation of whole population with different distribution

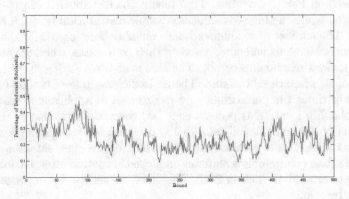

(c) The simulation of graduate school when prospective applicants with different distribution

Fig. 1. Result over 500 rounds

Table 1. Values of parameters used in the simulations

Number of both rich and high capacity applicant	N	40
Number of both poor and high capacity applicant	N	160
Number of both rich and low capacity applicant	N	160
Number of both poor and low capacity applicant	N	640
Number of graduate school's member	M	100
Type of applicant	K	4
Number of round	M	500
The current wage level	w_{l0}	100
The future wage level without postgraduate degree	w_{l1}	130
The future wage level with postgraduate degree	w_{h1}	162
Acquiring degree probability of candidate with high capacity	P_h	0.8
Acquiring degree probability of candidate with low capacity	P_l	0.5
Discount rate	β	0.1
The benchmark scholarship level	s	100

a low level of real scholarship, around 22 (see Fig. 1(c)), to block students from poor family, even with high capacity (Proposition 3). Thus, the strategy among different types of applicants tends to polarise, where people from rich family will apply and people from poor family will join to the job market(see Fig. 1(a)). The average mixed strategy of the whole population will converge to 0.2 as a steady state(see Fig. 1(b)).

5 Conclusion

This paper examines the behaviors of both prospective students and graduate schools. The graduate school will set a barrier of maximum scholarship to ensure the optimal payoff of graduate school. The decision of applying for the postgraduate program is determined by the current wage level for job applicants without postgraduate degree and the expected wage premium of a postgraduate degree. If the possibility of student acquiring excellent academic attainments is less than the ratio of postgraduate program's cost to value of a postgraduate degree, graduate schools will set up a maximum of scholarship to block students from poor family even though they have high capacity. A limitation of this study is absent of the heterogeneity analysis. Further research might explore decisions under different current wage level or expected wage premium of a postgraduate degree.

Acknowledgement. This work is supported in part by the scholarship from China Scholarship Council (CSC) under the Grant CSC NO.201606750015. We thank Settlement Co. (Waterloo) for its wonderful coffee and friendly staffs.

References

Brailsford, I.: Motives and aspirations for doctoral study: Career, personal, and interpersonal factors in the decision to embark on a history PhD. Int. J. Dr. Stud. **5**(1), 16–27 (2010)

Churchill, H., Sanders, T.: Getting your PhD: a practical insider's guide. Sage (2007)

Gill, T.G., Hoppe, U.: The business professional doctorate as an informing channel: a survey and analysis. Int. J. Dr. Stud. **4**(1), 27–57 (2009)

Golde, C.M.: The role of the department and discipline in doctoral attrition: lessons from four departments. J. High. Educ. **76**(6), 669–700 (2005)

Henrickson, L.: Old wine in a new wineskin: college choice, college access using agent-based modeling. Soc. Sci. Comput. Rev. **20**(4), 400–419 (2002)

Lindley, J., Machin, S.: The rising postgraduate wage premium. Economica **83**(330), 281–306 (2016)

Macal, C.M., North, M.J.: Agent-based modeling and simulation: ABMS examples. In: 2008 Winter Simulation Conference, WSC 2008, pp. 101–112. IEEE, December 2008

Mincer, J.: Investment in human capital and personal income distribution. J. Polit. Econ. **66**(4), 281–302 (1958)

Mincer, J.: Schooling, Experience, and Earnings. National Bureau of Economic Research, Cambridge (1974)

Nowak, M.A., Sasaki, A., Taylor, C., Fudenberg, D.: Emergence of cooperation and evolutionary stability in finite populations. Nature **428**(6983), 646–650 (2004)

Powell, S., Green, H. (eds.): The Doctorate Worldwide. Open University Press, Maidenhead (2007)

Reardon, S., Kasman, M., Klasik, D., Baker, R.: Agent-based simulation models of the college sorting process. J. Artif. Soc. Soc. Simul. **19**(1), 8 (2016)

Subgroup Optimal Decisions in Cost–Effectiveness Analysis

E. Moreno[1]([⊠]), F. J. Vázquez–Polo[2], M. A. Negrín[2], and M. Martel–Escobar[2]

[1] Departamento de Estadístca e I.O., Universidad de Granada, Granada, Spain
emoreno@ugr.es
[2] Dpto. de Métodos Cuantitativos, Universidad de Las Palmas de G.C.,
Las Palmas, Spain
{francisco.vazquezpolo,miguel.negrin,maria.martel}@ulpgc.es

Abstract. In cost–effectiveness analysis (CEA) of medical treatments the optimal treatment is chosen using an statistical model of the cost and effectiveness of the treatments, and data from patients under the treatments. Sometimes these data also include values of certain deterministic covariates of the patients which usually have valuable clinical information that would be incorporated into the statistical treatment selection procedure. This paper discusses the usual statistical models to undertake this task, and the main statistical problems it involves.

The consequence is that the optimal treatments are now given for patient subgroups instead of for the patient population, where the subgroup are defined by those patients that share some covariate values, for instance age, gender, etc. Some of the covariates are non necessarily influential, as typically occurs in regression analysis, and an statistical variable selection procedure is called for. A Bayesian variable selection procedure is presented, and optimal treatments for subgroups defined by the selected covariates are then found.

Keywords: Cost–effectiveness · Bayesian variable selection
Optimal treatments

1 Introduction

Given a set of alternative treatments $\{T_1, \ldots, T_k\}$ and the parametric class of distributions of their cost and effectiveness $\{P_i(c, e|\theta_i), i = 1, \ldots, k\}$, where θ_i is a unknown parameter, the reward of treatment T_i is the predictive distribution $P_i(z|R, \text{data}_i)$ of the net benefit $z = R \times e - c$ of the treatment, where R is the amount of money the decision maker is willing to pay for the unit of effectiveness and data$_i$ the available data of cost and effectiveness from patients under the treatment. Furthermore, if a utility function $U(z|R)$ of the net benefit z is assumed the cost–effectiveness analysis provides an optimal treatment for the whole population patient, conditional on R and the data sets $\{\text{data}_i, i = 1, \ldots, k\}$.

© Springer Nature Switzerland AG 2019
E. Bucciarelli et al. (Eds.): DCAI 2018, AISC 805, pp. 67–74, 2019.
https://doi.org/10.1007/978-3-319-99698-1_8

Additionally, let us suppose that jointly with the cost and effectiveness dataset a set of covariates $\mathbf{x} = (x_1, \ldots, x_p)$ of every patient under the treatment are available. The covariates indicate certain deterministic physical characteristics of the patients as age and sex, variables from their clinical history, or semiological variables of the disease [1]. This way the predictive distribution of the net benefit of treatment T_i will depend not only on R and data$_i$ but also on the covariates $\mathbf{x}$, that is, we will now have the rewards $P_i(z|R, \text{data}_i, \mathbf{x})$ for $i = 1, \ldots, k$. The adaptation of the cost–effectiveness analysis to this new situation yields the cost–effectiveness analysis for subgroups.

The aim of the subgroup analysis is that of finding optimal treatments with respect to a given utility function for patient subgroups defined by $\mathbf{x}$. It is apparent that for a given utility function $U(z|R)$ the expected utility of the reward $P_i(z|R, \text{data}_i, \mathbf{x})$,

$$\mathbb{E}_{P_i}(U) = \int_{\mathcal{Z}} U(z|R) P_i(z|R, \text{data}_i, \mathbf{x}) \, dz,$$

is a function of R, data$_i$ and $\mathbf{x}$. Thus, the dependency of $\mathbb{E}_{P_i}(U)$ on $\mathbf{x}$ implies that the resulting optimal treatment might change as $\mathbf{x}$ changes and then the optimal treatment for different subgroups might differ each other. The optimal treatment for the whole patient population might not be of interest.

Since the definition of patient subgroups is made in terms of the potential set of covariates $\mathbf{x}$ it is important to exclude from $\mathbf{x}$ those covariates that do not have an influence on the disease. This means that a previous step before carrying out a cost–effectiveness analysis for subgroups should be the early detection of the influential covariates from the original set of them. The statistical selection of covariates is known as the variable selection problem, an old and important problem in regression analysis for which many different methods have been utilized along its long history including those based on p–values (R^2, R^2 corrected, Mallows C_p), those based on Bayes factor approximations (Akaike AIC, Schwarz BIC, Spiegelhalter et al. DIC), or those based on Bayes factors (Zellner g– priors, Casella and Moreno intrinsic priors), among others.

The goal of this paper is the reformulation of the subgroup cost–effectiveness analysis to include variable selection in the regression models proposed, and the derivation of subgroup optimal treatments by applying the Decision Theory. We note that the statistical variable selection procedure typically yields not only a more realistic definition of the subgroups but also a desirable reduction of the dimension of the regression model.

We remark that the statistical variable selection introduces model uncertainty in the cost–effectiveness analysis, an uncertainty that should be incorporated into the decision making. The Bayesian methodology accounts for this uncertainty in an automatic way.

2   The Sampling Model

Let us assume that the data collected from n_i patients under treatment T_i, $i = 1, \ldots, k$, are given by the triple $(\mathbf{e}_i, \mathbf{c}_i, \mathbf{X})$, where $\mathbf{e}_i = (e_{i1}, \ldots, e_{in_i})^\top$ is a sample of the effectiveness of T_i, $\mathbf{c}_i = (c_{i1}, \ldots, c_{in_i})^\top$ a sample of the cost of T_i, and the design matrix

$$\mathbf{X} = \begin{pmatrix} 1 & x_{11} & \cdots & x_{p1} \\ \vdots & \vdots & \ddots & \vdots \\ 1 & x_{1n_i} & \cdots & x_{pn_i} \end{pmatrix}$$

is a matrix of dimensions $n_i \times (p+1)$, where the column $s+1$ contains the values $(x_{s1}, \ldots, x_{sn_i})^\top$ of the covariate x_s from patients under T_i, $s = 1, \ldots, p$.

The normal–normal model assumes that the cost and the effectiveness of treatment T_i, $i = 1, \ldots, k$, are continuous random variable that can be written as

$$c_i = \alpha_{i0} + \sum_{s=1}^{p} \alpha_{is} x_s + \alpha_{i,p+1}\, e_i + \varepsilon_i, \quad \text{and} \quad e_i = \beta_{i0} + \sum_{s=1}^{p} \beta_{is} x_s + \varepsilon_i', \quad (1)$$

where the random error ε_i and ε_i' follow the normal distribution $\mathcal{N}(\varepsilon_i|0, \sigma_i^2)$ and $\mathcal{N}(\varepsilon_i'|0, \tau_i^2)$, σ_i^2 and τ_i^2 being unknown variance errors, $\boldsymbol{\alpha}_i = (\alpha_{i0}, \alpha_{i1}, \ldots, \alpha_{ip}, \alpha_{i,p+1})^\top$ and $\boldsymbol{\beta}_i = (\beta_{i0}, \ldots, \beta_{ip})^\top$ are unknown regression coefficients.

We note that the cost model includes the effectiveness as a regressor. This simple form of conditioning the cost on the effectiveness has been considered by Willke et al. in [2], Willan and Kowgier in [3] and Moreno et al. in [4,5].

Therefore, the model for the sample cost $\mathbf{c}_i = (c_{i1}, \ldots, c_{in_i})^\top$ and effectiveness $\mathbf{e}_i = (e_{i1}, \ldots, e_{in_i})^\top$ can be written as

$$\mathcal{N}_{n_i}(\mathbf{c}_i | \mathbf{X}_c \boldsymbol{\alpha}_i, \sigma_i^2 \mathbf{I}_{n_i}) \times \mathcal{N}_{n_i}(\mathbf{e}_i | \mathbf{X}_e \boldsymbol{\beta}_i, \tau_i^2 \mathbf{I}_{n_i}), \quad (2)$$

where the design matrix $\mathbf{X}_c$ has dimensions $n_i \times (p+2)$ and is defined by adding to the design matrix $\mathbf{X}$ a column with the effectiveness $\mathbf{e}_i$, that is,

$$\mathbf{X}_c = \begin{pmatrix} 1 & x_{11} & \cdots & x_{p1} & e_{i1} \\ \vdots & \vdots & \ddots & \vdots & \vdots \\ 1 & x_{1n_i} & \cdots & x_{pn_i} & e_{in_i} \end{pmatrix},$$

$\mathbf{X}_e = \mathbf{X}$, and $\mathbf{I}_{n_i}$ is the identity matrix of dimensions $n_i \times n_i$.

3   Bayesian Variable Selection and Patient Subgroups

For the construction of the patient subgroups of a generic treatment T it is crucial the covariates we include in the regression model of the cost c and the effectiveness e. The question is whether p, the initial number of covariates, can be reduced by selecting a subset of them that have an influence on (c, e) based on the sampling information $\mathbf{c} = (c_1, \ldots, c_n)^\top$ and $\mathbf{e} = (e_1, \ldots, e_n)^\top$ from patients under treatment T. This is a decision problem with sampling information $(\mathbf{c}, \mathbf{e})$, where the decision space is the set of 2^p regression models $\mathfrak{M}$ for $\mathbf{c}$ and $\mathbf{e}$ defined by all possible subsets of the original set of p covariates. This set can be written as the union

$$\mathfrak{M} = \bigcup_{k=0}^{p} \mathfrak{M}_k$$

where $\mathfrak{M}_k$ is the set of regression models with k regressors. The reward of each model M_k in $\mathfrak{M}$ is their posterior probability conditional on the sample $\mathbf{y} = (\mathbf{c}, \mathbf{e})$. The utility function is a 0–1 function whose meaning is to win 1 unit when making the decision of choosing the true model for $\mathbf{y}$, and 0 otherwise.

It is immediate to see that the optimal model is the one having the highest posterior probability. Thus, the quantity of interest in variable selection is the posterior distribution of the models in $\mathfrak{M}$, conditional on the sample $\mathbf{y} = (\mathbf{c}, \mathbf{e})$ and the design matrices $\{\mathbf{X}_k, \ k = 1, \ldots, 2^p\}$.

Here we give a Bayesian procedure for variable selection in normal regression models for the hierarchical uniform prior distributions for models and the intrinsic priors for model parameters. Three reasons justify our selection: (1) are completely automatic, (2) the computation of the posterior model probability is simple and (3) present very good sample behavior.

3.1   Notation

For a generic treatment T, let $(\mathbf{y}, \mathbf{X})$ be the observed responses and covariates from n independent patients. We assume that $(\mathbf{y}, \mathbf{X})$ follows the n dimensional normal distribution $\mathcal{N}_n(\mathbf{y}|\mathbf{X}\boldsymbol{\beta}_{p+1}, \sigma_p^2 \mathbf{I}_n)$. This model that includes all covariates is called the full model, and is denoted as M_p. The model $\mathcal{N}_n(\mathbf{y}|\beta_0 \mathbf{1}_n, \sigma_0^2 \mathbf{I}_n)$ that includes no covariates is called *the intercept only model*, and is denoted as M_0.

By M_j we denote a generic model $\mathcal{N}_n(\mathbf{y}|\mathbf{X}_{j+1}\boldsymbol{\beta}_{j+1}, \sigma_j^2 \mathbf{I}_n)$ with j covariates, $0 \leq j \leq p$, where $\boldsymbol{\beta}_{j+1} = (\beta_0, \beta_1, \ldots, \beta_j)^\top$, $\mathbf{X}_{j+1}$ a $n \times (j+1)$ submatrix of $\mathbf{X}$ formed with j specific covariates, and σ_j^2 the variance error. The number of models M_j with j regressors is $p!/(j!(p-j)!)$, and the set of them is denoted as $\mathfrak{M}_j$. The class of all possible regression models with at most p regressors is $\mathfrak{M} = \cup_{j=0}^{p} \mathfrak{M}_j$. We remark that the regression coefficients change across models, although for simplicity we use the same alphabetical notation.

To complete the sampling models we need a prior for models in $\mathfrak{M}$, and for model parameters. Thus, for a generic model M_j we need a prior $\pi(\boldsymbol{\beta}_{j+1}, \sigma_j^2, M_j)$. It is convenient to decompose this prior as

$$\pi(\boldsymbol{\beta}_{j+1}, \sigma_j^2, M_j) = \pi(\boldsymbol{\beta}_{j+1}, \sigma_j^2 | M_j) \pi(M_j)$$

for $M_j \in \mathfrak{M}_j$, $(\boldsymbol{\beta}_{j+1}, \sigma_j^2) \in \mathbb{R}^{j+1} \times \mathbb{R}^+$.

3.2   Posterior Model Probability

Given the sample $(\mathbf{y}, \mathbf{X})$ from a model in the class $\mathfrak{M}$ it follows from the Bayes theorem that the posterior probability of model M_j, which contains j specific covariates, is given by

$$\Pr(M_j|\mathbf{y},\mathbf{X}) = \frac{m_j(\mathbf{y},\mathbf{X})\,\pi(M_j)}{\sum_{i=0}^{p}\sum_{M_i\in\mathfrak{M}_i} m_i(\mathbf{y},\mathbf{X})\,\pi(M_i)}, \tag{3}$$

where

$$m_i(\mathbf{y},\mathbf{X}) = \int \mathcal{N}_n(\mathbf{y}|\mathbf{X}_{i+1}\boldsymbol{\beta}_{i+1},\sigma_i^2\mathbf{I}_n)\pi(\boldsymbol{\beta}_{i+1},\sigma_i^2|M_i)\,d\boldsymbol{\beta}_{i+1}\,d\sigma_i^2$$

denotes the marginal distribution of the sample for model M_i.

If we divide the numerator and the denominator of (3) by the marginal of the data under the intercept only model $m_0(\mathbf{y},\mathbf{X})$, the posterior probability of model M_j becomes

$$\Pr(M_j|\mathbf{y},\mathbf{X}) = \frac{B_{j0}(\mathbf{y},\mathbf{X})\pi(M_j)}{\sum_{i=0}^{p}\sum_{M_i\in\mathfrak{M}_i} B_{i0}(\mathbf{y},\mathbf{X})\,\pi(M_i)}, \tag{4}$$

where

$$B_{i0}(\mathbf{y},\mathbf{X}) = \frac{m_i(\mathbf{y},\mathbf{X})}{m_0(\mathbf{y},\mathbf{X})}$$

is the Bayes factor for comparing models M_i and M_0. We note that the Bayes factor $B_{i0}(\mathbf{y},\mathbf{X})$ is nothing else but the ratio of the likelihood of model M_i and model M_0 for the data $(\mathbf{y},\mathbf{X})$. We also note that model M_0 is nested in model M_i.

The advantage of writing the posterior model probability as we do in (4) is that the Bayes factors there only involve nested models. It is known that the Bayes factor for nested models enjoy, under mild conditions, excellent asymptotic properties.

3.3   The Hierarchical Uniform Prior for Models

Since $\mathfrak{M}$ is a discrete space with 2^p models, a natural default prior over this space is the uniform prior, but, as [6] showed, it is not necessarily a good prior. We propose to use the following decomposition of the model probability

$$\pi^{\mathrm{HU}}(M_j) = \pi(M_j|\mathfrak{M}_j)\pi(\mathfrak{M}_j),$$

where

$$\pi(M_j) = \int_0^1 \theta^j(1-\theta)^{p-j}d\theta = \binom{p}{j}^{-1}\frac{1}{p+1}, \tag{5}$$

so that the model prior distribution, conditional on the class $\mathfrak{M}_j$, is uniform and the marginal prior on the classes $\mathfrak{M}_j$ is also uniform.

3.4   Intrinsic Priors for Model Parameters

Intrinsic priors for computing Bayes factors in variable selection have been used by many authors (see [6] and reference therein).

The standard intrinsic method for comparing the null model M_0 *versus* the alternative M_j provides the proper intrinsic prior for the parameters $(\boldsymbol{\beta}_{j+1}, \sigma_j)$ conditional on an arbitrary point (α_0, σ_0),

$$\pi^I(\boldsymbol{\beta}_{j+1}, \sigma_j | \alpha_0, \sigma_0) = \mathcal{N}_{j+1}(\boldsymbol{\beta}_{j+1} | \tilde{\boldsymbol{\alpha}}_0, (\sigma_j^2 + \sigma_0^2)\mathbf{W}_{j+1}^{-1}) \, HC^+(\sigma_j | \sigma_0),$$

where $\tilde{\boldsymbol{\alpha}}_0 = (\alpha_0, \mathbf{0}_j^\top)^\top$, $\mathbf{W}_{j+1}^{-1} = \dfrac{n}{j+2}(\mathbf{X}_j^\top \mathbf{X}_{j+1})^{-1}$, and

$$HC^+(\sigma_j | \sigma_0) = \frac{2}{\pi} \frac{\sigma_0}{\sigma_j^2 + \sigma_0^2}$$

is the half Cauchy distribution on $\mathbb{R}^+$ with location parameter 0 and scale σ_0. The unconditional intrinsic prior for $(\boldsymbol{\beta}_{j+1}, \sigma_j)$ is then given by

$$\pi^I(\boldsymbol{\beta}_{j+1}, \sigma_j) = \int \pi^I(\boldsymbol{\beta}_{j+1}, \sigma_j | \alpha_0, \sigma_0) \, \pi^N(\alpha_0, \sigma_0) \, d\alpha_0 \, d\sigma_0.$$

where $\pi^N(\alpha_0, \sigma_0) \propto 1/\sigma_0$ and the posterior distribution of $(\boldsymbol{\beta}, \tau^2)$, conditional on $(\mathbf{e}, \mathbf{X}_e)$, is

$$\pi(\boldsymbol{\beta}, \tau^2 | \mathbf{e}, \mathbf{X}_e) = \mathcal{N}_m \text{InvGa}(\boldsymbol{\beta}, \boldsymbol{\tau}^2 | \widehat{\boldsymbol{\beta}}, (\mathbf{X}_e' \mathbf{X}_e)^{-1}; \frac{\nu}{2}, \frac{\nu}{2} s_e^2), \tag{6}$$

where $\mathcal{N}_q \text{InvGa}$ represents a q–variate Normal–inverted–Gamma distribution, $\widehat{\boldsymbol{\alpha}} = (\mathbf{X}_c^\top \mathbf{X}_c)^{-1}\mathbf{X}_c^\top \mathbf{c}$ and $\widehat{\boldsymbol{\beta}} = (\mathbf{X}_e^\top \mathbf{X}_e)^{-1}\mathbf{X}_e^\top \mathbf{e}$ are the MLE estimators of $\boldsymbol{\alpha}$ and $\boldsymbol{\beta}$, respectively, $\nu' = n - m'$ and $\nu = n - m$ are degrees of freedom, $s_c^2 = RSS_c/\nu'$ and $s_e^2 = RSS_e/\nu$ are the usual unbiased estimators of the variances σ^2 and τ^2, respectively, and $RSS_c = (\mathbf{c} - \mathbf{X}_c\boldsymbol{\alpha})^\top(\mathbf{c} - \mathbf{X}_c\boldsymbol{\alpha})$ and $RSS_e = (\mathbf{e} - \mathbf{X}_e\boldsymbol{\beta})^\top(\mathbf{e} - \mathbf{X}_e\boldsymbol{\beta})$.

Then, as in Moreno et al. (2018) (see [7]), we can derive the posterior predictive distribution of the effectiveness e, given the data, as follows. Let $\mathbf{x}_e$ be a generic vector of regressors for the effectiveness e. Then, the distribution of e is

$$f(e | \boldsymbol{\beta}, \tau^2, \mathbf{x}_e) = \mathcal{N}(e | \mathbf{x}_e^\top \boldsymbol{\beta}, \tau^2)$$

and from the properties of the normal–inverted–gamma distribution, the posterior predictive marginal distribution of e is the following Student t distribution

$$f(e | \mathbf{x}_e) = t(e | \mathbf{x}_e \widehat{\boldsymbol{\beta}}, (1 + \mathbf{x}_e(\mathbf{X}_e^\top \mathbf{X}_e)^{-1}\mathbf{x}_e^\top)s_e^2; \nu). \tag{7}$$

Analogously, from the conditional distribution of c given e and a generic regressor $\mathbf{x}_c(e)$ is

$$f(c | e, \boldsymbol{\alpha}, \sigma^2, \mathbf{x}_c(e)) = \mathcal{N}(c | \mathbf{x}_c(e)\boldsymbol{\alpha}, \sigma^2).$$

The posterior predictive of the cost c, conditional on e, the data and the regressors $\mathbf{x}_c(e)$ is the following Student t distribution

$$f(c|e, \mathbf{x}_c(e)) = t(c|\widehat{\boldsymbol{\alpha}}^\top \mathbf{x}_c(e), (1 + \mathbf{x}_c(e)(\mathbf{X}_c^\top \mathbf{X}_c)^{-1}\mathbf{x}_c^\top(e))s_c^2; \nu'). \tag{8}$$

Thus, from Eqs. (7) and (8) we have that the joint predictive distribution of c and e, conditional on the data and the generic regressors $\mathbf{x}_e$ and $\mathbf{x}_c(e)$, is

$$f(c, e|\mathbf{x}_c(e), \mathbf{x}_e) = t(c|\widehat{\boldsymbol{\alpha}}^\top \mathbf{x}_c(e), (1 + \mathbf{x}_c(e)(\mathbf{X}_c'\mathbf{X}_c)^{-1}\mathbf{x}_c^\top(e))s_c^2; \nu')$$
$$\times\, t(e|\widehat{\boldsymbol{\beta}}^\top \mathbf{x}_e, (1 + \mathbf{x}_e(\mathbf{X}_e^\top \mathbf{X}_e)^{-1}\mathbf{x}_e^\top)s_e^2; \nu). \tag{9}$$

Unfortunately, $f(z|R, \text{data}, \mathbf{X}_e, \mathbf{X}_c, \mathbf{x}_e, \mathbf{x}_c)$, the predictive distribution of the net benefit z has no known explicit distribution. However, sampling from the predictive distribution of the net benefit z is straightforward: first for each R and the covariate $\mathbf{x}_e$ we sample from $f(e|\mathbf{x}_e)$ in Eq. (7), and then for the covariate $\mathbf{x}_c$ from $f(c|e, \mathbf{x}_c)$ in Eq. (8). From these samples of e's and c's we immediately have a sample of the net benefit z.

4   Optimal Treatments for Subgroups

For simplicity in the presentation, we consider the particular case of two treatments T_1 and T_2 and denote z_1 the net benefit of treatment T_1 and z_2 the net benefit of treatment T_2. For the utility function $U_1(z_i|R) = z_i$, and a fixed value of R, the set of covariates values for which treatment T_1 is optimal is given by

$$\mathfrak{C}_R^{U_1} = \left\{\mathbf{x} : \varphi_R(\mathbf{x}) \geq 0\right\},$$

where

$$\varphi_R(\mathbf{x}) = \mathbb{E}_{P_1}(z_1|R, \text{data}_1, \mathbf{x}) - \mathbb{E}_{P_2}(z_2|R, \text{data}_2, \mathbf{x})$$

The computational difficulty for characterizing $\mathfrak{C}_R^{U_1}$ depends on the complexity of the function $\varphi_R(\mathbf{x})$. When the components of $\mathbf{x}$ are discrete, the set $\mathfrak{C}_R^{U_1}$ is easily characterized by direct evaluation of the functions $\varphi_R(\mathbf{x})$.

5   Discussion and Conclusions

In the presence of covariates a cost–effectiveness analysis to determine subgroup optimal treatments has been proposed. The procedure is a statistical decision problem that we developed using the Bayesian approach. We argued that non influential covariates should be deleted from the models as they are crucial for defining subgroups. Variable selection is a model selection problem, and we applied a Bayesian variable selection procedure that utilizes the intrinsic priors for the model parameters and the hierarchical uniform prior for the models. This procedure has shown to enjoy excellent statistical properties. Further, the cost and effectiveness are not necessarily independent random variables, and the Bayesian variable selection procedure indicates whether or not should we include the effectiveness as a covariate in the regression model for the cost.

The developments have been given for the normal–normal and the lognormal–normal regression models for the cost and effectiveness. When the effectiveness is measured by a discrete 0–1 random variable a Logistic or Probit model can be an appropriated model. Unfortunately, the Bayesian variable selection procedure for these model is far from simple, and variable selection is still an open problem. An involved solution for Probit models has been proposed in [8] although the topic still deserves much more research efforts.

Acknowledgments. Research partially supported by Grants ECO2017–85577–P (MINECO, Spain) and CEICANARIAS2017–025 (Canary Islands).

References

1. Sculpher, M., Gafni, A.: Recognizing diversity in public preferences: the use of preference sub-groups in cost-effectiveness analysis. Health Econ. **10**(4), 317–324 (2001)
2. Willke, R., Glick, H.A., Polsky, D., Schulman, K.: Estimating country-specific cost-effectiveness from multinational clinical trials. Health Econ. **7**, 481–493 (1998)
3. Willan, A.R., Kowgier, M.E.: Cost-effectiveness analysis of a multinational RCT with a binary measure of effectiveness and an interacting covariate. Health Econ. **17**(7), 777–791 (2008)
4. Moreno, E., Girón, F.J., Vázquez-Polo, F.J., Negrín, M.A.: Optimal healthcare decisions: the importance of the covariates in cost-effectiveness analysis. Eur. J. Oper. Res. **218**(2), 512–522 (2012)
5. Moreno, E., Girón, F.J., Martínez, M.L., Vázquez-Polo, F.J., Negrín, M.A.: Optimal treatments in cost-effectiveness analysis in the presence of covariates: improving patient subgroup definition. Eur. J. Oper. Res. **226**(1), 173–182 (2013)
6. Moreno, E., Girón, J., Casella, G.: Posterior model consistency in variable selection as the model dimension grows. Stat. Sci. **30**(2), 228–241 (2015)
7. Moreno, E., Vázquez-Polo, F.J., Negrín, M.A.: Cost-Effectiveness Analysis of Medical Treatments: A Statistical Decision Theory Approach. Chapman & Hall/CRC Biostatistics Series, Boca Raton (2018, in press)
8. León-Novelo, L., Moreno, E., Casella, G.: Objective Bayes model selection in probit models. Stat. Med. **31**(4), 353–365 (2012)

A Statistical Tool as a Decision Support in Enterprise Financial Crisis

Francesco De Luca[1]([⊠]), Stefania Fensore[2], and Enrica Meschieri[1]

[1] DEA, University "G. d'Annunzio" of Chieti-Pescara,
Viale Pindaro 42, 65127 Pescara, Italy
francesco.deluca@unich.it
[2] DSFPEQ, University "G. d'Annunzio" of Chieti-Pescara,
Viale Pindaro 42, 65127 Pescara, Italy

Abstract. The recent reform of Italian Insolvency Law introduced new instruments aimed to restore companies in financial distress and potential in bankruptcy. In particular, the Article 182-bis restructuring agreements has been introduced by the Italian Civil Code to manage company crisis. The objective of this study is to underline the ability of seven specific accounting ratios and coefficients to predict the status of financial distress of the firms. We introduce a new formula that we call M-Index indicator, and then provide an empirical analysis through a sample of Italian listed companies collected from the Milan Stock Exchange in the period 2003–2012. The results of the empirical analysis validate the predictive accuracy power of our indicator.

Keywords: Troubled debt restructuring (TDR) · Accounting ratios
Financial distress · Prediction

1 Introduction

The global economic Italian crisis dating back to 2009 led to the presence of a growing number of companies experiencing a financial distress. The urgent need of these companies to restructure their debt brought the Italian Law to introduce the troubled debt restructuring (TDR) procedure, through the Article 182-bis restructuring agreements (see [6]). The restructuring agreement is defined "as an operation whereby the creditor (or a group of creditors) grants a concession to the debtor in financial difficulties, such that otherwise it would not have agreed" (OIC 6, July 2011), see [11].

This new approach is a pre-insolvency proceeding quite similar to Chapter 11 of the U.S. Bankruptcy Code (see [1]). However, some differences arise. The main difference is represented by the "cram down" rule. In the U.S. legislation, if a class of creditors vote in favor of the agreement versus one (or more) classes, the restructuring plan presented by the debtor can be approved. On the contrary, according to Article 182-bis, the Italian Court proceeds with formal supervision to verify the restructuring plan approval. Therefore, the purpose of Article 182-bis is to restore companies that come across a financial distress, but the success of this procedure depends on the timeliness of the intervention.

© Springer Nature Switzerland AG 2019
E. Bucciarelli et al. (Eds.): DCAI 2018, AISC 805, pp. 75–82, 2019.
https://doi.org/10.1007/978-3-319-99698-1_9

Only few authors [7, 10] classified "failed firms" as companies that file for bankruptcy, including Chapter 11 and all bankruptcy tools. Accordingly, Chapter 11 and bankruptcy are considered similar. But, from a legal point of view, there is a distinction from the U.S. legislation (Chapter 7 and Chapter 11) and the Italian Insolvency Law (Article 182-bis and Article 216 et seq.). Note that Chapter 7 is devoted only to bankruptcy, while Chapter 11 deals with an agreement between debtors and creditors to restore a financial distressed firm.

In the TDR literature, there are a number of definitions of "financial distress". Several authors affirm that companies that decide to access the TDR agreement are in a status of financial difficulty, but this situation is different from bankruptcy. However, the boundaries of such a difference are not entirely clear (see, for example, [2, 4, 10, 12]). According to [10, 12], financial distress consists of acting to file for bankruptcy, i.e. a "filed firm" is a company that must file for bankruptcy. These authors consider the conditions of financial distress and failure as similar events. On the contrary, in this work we conceive the financial distress as something completely different from bankruptcy.

In order to predict the status of financial distress, the most famous and well known tool is the Altman's Z-score model [2], that has been deeply studied within the literature. Although some researches (see, for example, [8]) confirmed the validity of this model and its predictive power, other studies (see, for example, [5, 9]) raised some doubts about its potential, highlighting the need to develop new approaches.

The availability of a specific tool to predict and forecast company distress or bankruptcy is generally important in decision making process for several reasons. If we consider the company perspective, in fact, we believe that companies' management could appreciate the possibility to understand as earlier as possible if the company is experiencing or approaching a phase of financial distress, especially when performance indicators are not yet apparently negative. This would help managers in taking sudden decisions to change their own strategic behavior to overcome the distress and avoid bankruptcy or liquidation. Moreover, in the external stakeholders perspective, the same tool could be very useful for capital providers, lenders and creditors in understanding if the company they are assessing is presenting (or not) some signals of a forthcoming crisis (even if it is not apparently visible). To this end, we introduce a new indicator, the M-Index indicator as an early warning decisional tool showing a considerable predictive power. It could reveal a very useful tool for Italian business companies in order to assess their financial distress status and address the following decisional process of all involved parties.

This chapter is organized as follows: Sect. 2 contains the description of the sample data and the methodology used to conduct the empirical analysis. Specifically, we describe how the sample has been collected and the accounting ratios selected. In Sect. 3 we report the outcomes of the analysis, and, as expected, we will find that they support the theoretical results. Then, in Sect. 4, we provide an analysis of the effectiveness of TDR on distressed companies by assessing companies' accounting ratios in the aftermath of the TDR request. Finally, Sect. 5 summarizes the content of the work with a brief conclusion.

2 Data Analysis

2.1 Sample Data

The data used to process the empirical analysis have been obtained from Aida Database (Bureau Van Dijk) and the Milan Stock Exchange list. The dataset consists of a panel data of 50 Italian companies observed from 2003 to 2012. The time period has been selected to consider at least two years of observation before the law reform was issued (2005), and seven subsequent years, in order to observe the companies' trend after the reform, given a reasonable time to perform (or not) improvements of their financial equilibrium. During the selected time period, only 20 Italian listed companies filed for TDR procedure and we considered all of them. Moreover, in order to create a control group, we selected 30 listed companies (that did not filed for TDR procedure) by following a stratified sampling scheme by size and industry. The ratio control/case of the sample sizes is traditionally equal to one, however we have prudentially set it to 1.5 in order to take into account the sample variability. Specifically, the sample refers to seven macro industries, as reported by the Milan Stock Exchange. In particular, the considered industry types are: Consumer goods, Consumer service, Finance, Health and Food, Manufacturing, Oil and Natural Gas, and Technology. The companies have been labelled as "distressed" and "non-distressed" according to if they have filed for Article 182-bis restructuring agreements or not. A representation of this data is reported in Fig. 1.

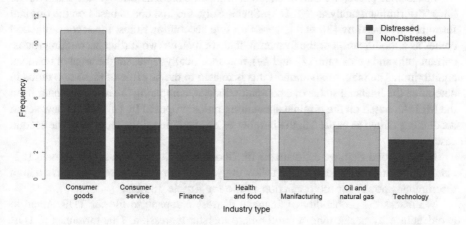

Fig. 1. Frequency distribution of the sample. Firms are classified according to industry type and "distressed/non-distressed" status.

In this study we consider specific accounting ratios, namely X_i, with $i = 1, ..., 7$, computed for each firm from 2003 to 2012. First, we consider the five ratios (X_1 to X_5) suggested by [2, 3], which are accounting ratios specific to liquidation/failure studies. Then, we decided to run the discriminant analysis by considering in addition the current ratio (X_6) and the quick ratio (X_7) because we believe that financial distress is signaled in advance by these two ratios. Specifically, the company's general equilibrium is based on both economic equilibrium (based on profitability) and financial equilibrium (based on liquidity and solvency). Therefore we decided to include also the two aforementioned specific indicators, X_6 and X_7, that are unanimously considered as liquidity and solvency ratios since they measure the ability of a company to meet short term cash needs. In this sense, once first signals of financial distress arise, a sudden intervention could prevent the deterioration of the economic equilibrium. Specifically, we have:

- X_1 working capital/total assets;
- X_2 retained earnings/total assets;
- X_3 earnings before interest and taxes/total assets;
- X_4 market value of equity/book value of total liabilities;
- X_5 sales/total assets;
- X_6 current assets and short-term liabilities;
- X_7 total liquidity and short-term liabilities.

Note that the longitudinal nature of our approach is due to the fact that the TDR procedure is viewed like an ongoing process and not a singular event.

2.2   The Methodology

The statistical analysis consists of two steps.

In the first phase, we derive two discriminant functions through the multivariate linear discriminant analysis (MLDA). Specifically, the first one is based on the original formula introduced by [2], that is based on five accounting ratios; the second function comes as a modification of the former, in that we include two further accounting ratios: current ratio and quick ratio (X_6 and X_7), that specifically represent the level of financial equilibrium. The main motivation of this is related to the key role of these two ratios in determine the financial status of a company. Indeed, comparing the results coming from the MLDAs based on the original accounting ratios proposed by [2] and the new seven accounting ratios, as expected, the number of misclassified firms reduces in the second case.

In the second step, we estimate the likelihood of each firm resorting to Article 182-bis restructuring agreements in the following year. In our case, these ratios have a high discriminant power for the prediction to file for Article 182-bis.

We assess the probability of financial distress company to file for TDR aimed to avoid failure by performing a multivariate logistic regression. Our formula for TDR probability is

$$\hat{P}_{it} = \frac{e^{S_{it-1}}}{1 + e^{S_{it-1}}}$$

where the logit S_{it} is a score associated to the considered accounting ratios for the i-th firm at the t-th period, and it is defined as

$$S_{it}\, score = 0.003X_{1it} + 0.267X_{2it} + 0.663X_{3it} + 0.431X_{4it} + 0.533X_{5it}$$
$$+ 0.147X_{6it} + 0.092X_{7it}$$

The estimation of the coefficients of the $S_{it}score$ comes from a MLDA where the ratios used for each firm correspond to: the ratios belonging to the year when the TDR was requested for distressed firms, and the best ratios observed during the considered years for non-distressed firms.

3 Results

The results of this analysis show that for distressed companies the probability to file for a TDR procedure should increase until the year of TDR request. When the companies files for a TDR, the probability becomes stationary (or decreases). On the other hand, for non-distressed companies the time series appear stationary with a constant trend in the period.

We also introduce a new indicator, called M-Index, in order to determine whether an increment of TDR probability is near to the request of this procedure. For the i-th firm, the M-Index is

$$M_i = \frac{\max(\hat{P}_{i,t_0}, \hat{P}_{i,t_0-1}, \hat{P}_{i,t_0-2})}{\sum_{t=t\,min}^{t\,max} \hat{P}_{i,t}/T}$$

The numerator is calculated as the maximum probability associated with the two years before the TDR request and the year when TDR in approved (t_0); t_{min} and t_{max} represent the first and the last observation year. In order to interpret this indicator we have its value as: greater than one for distressed companies before the TDR request, less than one for non-distressed firms.

The M-Index indicator has revealed its strong potential when predicting the TDR probabilities for the sample described in Sect. 2.1. From the empirical analysis we obtained the following results: the M-Index value was, as expected, lower than one for all the firms belonging to the non-distressed group; while, for the other group, we obtained an M-Index value almost equal to one in six cases and greater than one in the remaining fourteen cases.

4 The Effectiveness of TDR: Main Results

As above mentioned, the TDR was introduced in the Italian Law in 2005, therefore, as far as we know, there are no previous analyses in the literature addressed to show the effectiveness of this quite new procedure. For this reason, we tried also to assess the effectiveness of the TDR in the period right after the homologation by the court. In this analysis we consider a positive impact if we observe both a decrease of the probability rate and a stand-still trend of it.

To this end, we separate the firms of the distressed group for each observation year from 2006 to 2011. Each group is labelled according to the year of the TDR request. We achieve six groups with the following labels: *"2006"*, *"2007"*, *"2008"*, *"2009"*, *"2010"*, *"2011"*. Companies that filed for TDR on 2012 are excluded from this analysis because too short time passed by to show a reliable impact of TDR on financial equilibrium. In fact, we preferred to consider only companies that had the possibility to spend at least two periods after the court approval.

Therefore, starting from the period t_0, each group has a number of observation periods as many as are the years to be observed. In summary, we have six observation periods for group "2006" and only one observation period for group "2011".

In the group "2006" there is a reduction of the probability rate to file the agreement. This is a signal of an evident improvement of accounting and financial ratios. Furthermore, it shows how the financial distress status stands still during the first three years after the validation of the TDR. In Fig. 2, we show the trends of the averages of TDR probabilities, and the progress of group "2006" is clear. This group is the only one that offers a more complete view of the effects of TDR on distressed companies, given that we observed five years after the homologation by the court.

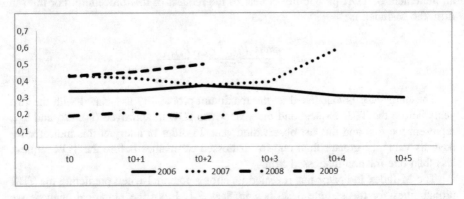

Fig. 2. Trends of average TDR probabilities after the request. Each average has been calculated among firms who filed for the TDR request in the same year. The request year is denoted by t_0.

The group "2007" shows an improvement over three periods after the approval of the agreement. The only exception is due to period t_{0+4}, where a worsening of economic and financial ratios appears. However, we posit that TDR impacted positively on distressed companies and the decrease of the last period of observation is due to an increase of requests of TDR from further companies different from the ones that filed in 2007.

For the group "2008" we can observe an improvement in the period t_{0+2} as well, with a consequent departure from the financial distress status. Instead, the group "2009" shows an increase of the probability rate. However, due to the short time after the homologation of TDR, it is not possible to assess subsequent improvement or worsening of the business situation for the groups "2010" and "2011".

5 Conclusions

In this research we introduced a statistical tool based on specific accounting and financial ratios with the aim to provide a decision tool to file for a TDR procedure when a firm is experiencing a distressed condition. We carried out an empirical analysis collecting a longitudinal dataset, ranging from 2003 to 2012, including twenty listed Italian companies that have filed the Article 182-bis restructuring agreement. Then, we collected from the Milan Stock Exchange thirty companies, labelled as "healthy", as the control group.

Through the MLDA, we created a simple function employed as the score formula in the predictive model and, then, we introduced a new indicator, called M-Index, to describe the economic and financial equilibrium using historical data, and forecast the financially distressed firm's probability of filing for TDR. Our results confirm the power of this tool in predicting the company's financial distress status across the different considered industries.

Moreover, we also tried to assess the benefit power of TDR to overcome financial distress and we found that the TDR has usually improved the health of a firm. In this sense, TDR could be an effective tool for companies experiencing financial difficulties and it has avoided their failure. By the way, even in the case TDR did not produce any improvement of financial equilibrium, it has been able to freeze the status of the distressed company. If we consider that this data refers to the period of the deepest financial crisis which has taken place in Italy over the past fifty years, and, for this reason, the external conditions for the recovery have been totally unfavourable, then we could express, with a mildly favourable opinion, about the effectiveness of TDR request.

Finally, this study aims to underline that the relevance of this predictive instrument is addressed not only to companies in financial difficulties but also to creditors, investors and banks, which can use this tool at the right time in order to early warn companies and suggest to them to face a debt restructuring and avoid failure when it is still possible.

Limitations of this study may lie on the sample size that is limited because in the Italian Stock Exchange only 20 companies accessed the TDR procedure during the years right after the reform. A further research could be addressed in order to enlarge the sample size and assess, along a wider time period, the efficiency of the TDR procedure in pursuing the aims that the legislator assigned to the reform of the bankruptcy law, i.e. the reduction of the rate of companies' failure during financial crisis period.

References

1. Altman, E.I., Danovi, A., Falini, A.: Z-score models' application to Italian companies subject to extraordinary administration. J. Appl. Financ. **23**(1), 128–137 (2013)
2. Altman, E.I.: Corporate Financial Distress and Bankruptcy: A Complete Guide to Predicting and Avoiding Distress and Profiting from Bankruptcy. Wiley, New York (1993)
3. Altman, E.I.: Financial ratios, discriminant analysis and the prediction of corporate bankruptcy. J. Financ. **23**, 589–609 (1968)

4. Beaver, W.H.: Financial ratios as predictors of failures. J. Account. Res. **4**, 71–111 (1966)
5. Begley, J., Ming, J., Watts, S.: Bankruptcy classification errors in the 1980s: an empirical analysis of Altman's and Ohlson's model. Rev. Account. Stud. **1**, 267–284 (1996)
6. Di Marzio, F.: Il nuovo diritto della crisi di impresa e del fallimento. ITA Edizioni, Torino (2006)
7. Gilson, S.C., John, K., Lang, L.H.P.: Troubled debt restructurings: an empirical analysis of private reorganization of firms in default. J. Financ. Econ. **27**, 315–353 (1990)
8. Hayes, S.K., Hodge, K.A., Hughes, L.W.: A study of the efficacy of Altman's Z to predict bankruptcy of specialty retail firms doing business in contemporary times. Econ. Bus. J. Inq. Perspect. **3**(1), 122–134 (2010)
9. Lifschutz, S., Jacobi, A.: Predicting bankruptcy: evidence from Israel. Int. J. Bus. Manag. **4** (5), 133–141 (2010)
10. Ohlson, J.A.: Financial ratios and probabilistic prediction of bankruptcy. J. Account. Res. **18**, 109–131 (1980)
11. Organismo Italiano di Contabilità (OIC). Ristrutturazione del debito e informativa di bilancio (Debt restructuring and disclosure). OIC 6 July 2011. http://bit.ly/2hE99vm. (in Italian)
12. Zmijewki, M.E.: Methodological issues related to the estimation of financial distress prediction models. J. Account. Res. **22**(Suppl.), 58–82 (1984)

A Mediation Model of Absorptive and Innovative Capacities: The Case of Spanish Family Businesses

Felipe Hernández-Perlines[(⊠)] (iD) and Wenkai Xu

University of Castilla-La Mancha,
Cobertizo de San Pedro Mártir, s/n, 45071 Toledo, Spain
Felipe.HPerlines@uclm.es, 349766649@qq.com

Abstract. This work analyses the mediating effect of innovative capacity on the influence of absorptive capacity in the performance of family businesses. For the analysis of results, the use of a second-generation structural equation method is proposed (PLS-SEM) using smartPLS 3.2.7 computer software, applied to the data of 218 Spanish family businesses. The main contribution of this work is that the performance of family businesses is determined by the absorptive capacity (absorptive capacity is able to explain 36.1% of the performance variability of family businesses). The second contribution of this work is that the influence of the absorptive capacity on the performance of family businesses is strengthened by the effect of innovative capacity, explaining 40.6% of the variability. The third contribution is that the absorptive capacity is a precedent for innovation capacity, able to explain 50% of its variability.

Keywords: Absorptive capacity · Innovative capacity · Family performance
Mediating effect · PLS

1 Introduction

The competitiveness of businesses depends on them renewing and adapting their resources and capacities in an environment subject to great changes and with high levels of uncertainty. This paper analyzes the absorptive and innovative capacity in de family business. Absorptive capacity is related with the identification, assimilation and exploitation of new knowledge [1]. The success of the business depends on its skill in recognising, assimilating and applying new knowledge [2]. Absorptive capacity, like dynamic capacity [3, 4] have under-gone various reformulations. The most relevant was carried out by Zahra and George [5] in 2002. For these authors, absorptive capacity is a set of organizational routines and processes by which firms acquire, assimilate, transform and exploit knowledge [5].

The objective of this paper is to analyze the mediating effect of innovative capacity on the influence of absorptive capacity on the performance of family businesses.

© Springer Nature Switzerland AG 2019
E. Bucciarelli et al. (Eds.): DCAI 2018, AISC 805, pp. 83–90, 2019.
https://doi.org/10.1007/978-3-319-99698-1_10

The companies in this study are family businesses located in Spain. The justification for this selection lies in the importance of this kind of business; they represent 89% of the businesses that operate in Spain, 57% of GDP, and 67% of private employment [6].

Therefore, this type of business is an important driver of economic growth and wellbeing [7].

To analyse the results and contrast the hypotheses, with a PLS-SEM model of structural equations, using the SmartPLS 3.2.7 computer program [8]. The data was obtained from a questionnaire sent by email to the CEOs of family businesses associated with the Family Business Institute during the months of June to November 2016. At the end of the process 216 completed questionnaires were obtained.

2   Theory and Hypothesis

The theoretical framework upon which this research is based is that of dynamic capacities [9–13] as both absorptive capacity and innovative capacity allow the business to adapt to the changing conditions of the environment.

Absorptive capacity arises as an essential research issue in business strategy [2]. Absorptive capacity allows businesses to survive certain pressures, recognising, assimilating and applying new knowledge [2]. The notion of absorptive capacity was originally developed by Cohen and Levinthal [1]. For these authors, absorptive capacity is the capacity of the business to identify, assimilate and exploit new knowledge.

Innovative capacity effectively links the inherent innovation of a business with the advantage based on the market in terms of new products and/or markets. Therefore, innovative capacity explains the links between re-sources and the capacities of a business with its market [14] Innovative capacity, according to Wang and Ahmed [14], refers to a business' capacity for developing new products and/or markets, through the alignment of strategic innovative orientation with innovative behaviours and processes.

The designed model poses two hypotheses:

H1: Absorptive capacity positively affects the performance of family businesses.
H2: The innovative capacity positively mediates the relationship be-tween absorptive capacity and performance of family businesses.

3   Methodology

3.1   Data

The data obtained from a questionnaire sent by post through the Lime-survey v. 2.5. tool to the CEO/Director of a sample of businesses taken from the Family Business Institute. The information about field work can be seen in Table 1.

Table 1. Technical datasheet of the fieldwork

Target population (universe)	1,054 Spanish family business
Analysis unit/Sampling unit	The company
Sample size/Response rate	218 valid surveys/20.86%
Confidence level	95%; z = 1.96; p = q=0.50; α = 0.05
Sampling error	5.91%
Key informant	CEO/Director
Date of fieldwork/Data collection	June–November 2016

3.2 Variables

Absorptive Capacity. To measure the absorptive capacity, the scale proposed by Cohen and Levinthal [1] and Lane et al. [15] and validated by Flatten et al. [16] has been considered. With this second order composite, the extent to which a company is dedicated to the acquirement of knowledge (3 items), assimilation (4 items), transformation (4 items) and exploitation (3 items) is evaluated.

Innovative Capacity. To measure the innovative capacity, the scale proposed by Prajogo and Sohal [17] has been considered. This second order composite is applied to two types of innovation: product innovation (measured from 5 items) and process innovation (4 items or indicators).

Performance. In this research, we measure business performance based on the scale proposed by Wiklund and Shepherd [18] and Chirico et al. [19] made up of 4 items. The original indicators of the different composites considered have been adapted to a Likert scale (1–7). The validity and reliability of the different composites considered in this work has been demonstrated in previous works such as Yáñez-Araque et al. [20] and Hernández-Perlines and Xu [21].

Control Variables. For control variables, size (number of employees), age (number of years since establishment) and the main activity sector of the family business were used, appearing on a recurring basis in studies on family businesses [22].

PLS-SEM was used to test the hypotheses and analyze the mediating effect of innovative capacity. SmartPLS v.3.2.7 was employed for the analysis [8].

4 Results

To interpret and analyse the model proposed in the PLS-SEM, two stages were developed [23]: (1) measurement model analysis; and (2) structural model analysis. This sequence assures the scales proposed for measurement are valid and reliable.

4.1   Measurement Model Analysis

Table 1 shows the parameters associated to the assessment of the measurement model. All the values of load factors exceed 0.5, which is considered as an acceptable level by Barclay et al. [23], and Chin [24].

The values of composite reliability and average variance extracted (AVE) exceed the 0.7 and 0.5 recommended limits, respectively [25]. The obtained values support the convergent validity of the considered scales. Finally, to guarantee discriminant validity, the correlations between each pair of constructs was meant not to exceed the square root of each construct's AVE (see Table 2).

Table 2. Correlation, composite reliability, convergent and discriminant validity

Composite/Measures	AVE	Composite reliability	1. ACAP	2. ICAP	3. FIPERF
1. Absorptive capacity (ACAP)	0.835	0.953	0.913*	0	0
2. Innovative capacity (ICAP)	0.893	0.957	0.710	0.944*	0
3. Firm performance (FIPERF)	0.717	0.910	0.657	0.645	0.846*

Note: Correlations are from the second-order CFA output.
(*) - The diagonal elements are the square root of the AVE.

In addition, the HTMT index for composites type a that allow the measurement of the discriminant validity between indicators of the same composite and between indicators of different composites. To fulfil discriminant validity, the HTMT ratio values must be less than 0.85 [26] (see Table 3).

Table 3. Ratio heterotrait-monotrait (HTMT) and descriptive statistics

Ratio heterotrait-monotrait (HTMT)	1. ACAP	2. ICAP	3. FIPERF
1. Absorptive capacity (ACAP)			
2. Innovative capacity (ICAP)	0.779		
3. Firm performance (FIPERF)	0.339	0.272	
Cronbach's Alpha	0.934	0.887	0.867
Mean	4.353	4.431	4.745
Typical deviation	0.946	1.125	1.667

4.2   Structural Model Analysis

To determine the different effects, we followed the steps proposed by Hair et al. [27] in order to apply the focus of Preacher and Hayes [28] to the proposed mediation model. Firstly, the direct effect is analyzed between the absorptive capacity and performance of the family businesses. To do so, it verifies the path coefficient value and its significance (it applies the bootstrapping procedure of 5000 subsamples). The effect is positive and significant ($\beta = 0.390$; $p < 0.001$) (see Fig. 1). Furthermore, the absorptive capacity can explain 36.1% of the performance variability of family businesses.

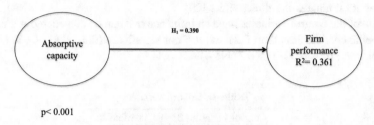

Fig. 1. Direct model.

The second step consists of including the effect of the mediating variable (absorptive capacity). In Fig. 2, it is observed that the indirect effect is positive and significant (between the absorptive capacity and innovative capacity H2a: $\beta = 0.707$; $p < 0.001$; and between the innovative capacity and the performance of family businesses H2b: $\beta = 0.358$; $p < 0.001$). The mediating effect fully eliminates the direct effect, as the direct relationship between the absorptive capacity and performance of the family business is $\beta = 0.046$ and is not significant.

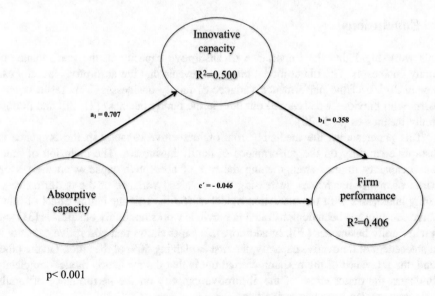

Fig. 2. Mediating model.

In the model, the absorptive capacity can explain 50% of the variability of innovative capacity and can in turn explain 40.6% of performance variability of family businesses. Therefore, mediation of innovative capacity is produced in relation to absorptive capacity and the performance of family businesses, and is a total mediating effect, as it eliminates the direct effect [29].

None of the control variables have an influence that can be considered relevant (the path coefficients are less than 0.2) and are not significant (the value of t is less than recommended (p < 0.001) (see Table 4).

Table 4. Control variables

Control variable	ß	t-value
Age	−0.011	0.089
Sector	−0.051	0.750
Size	0.121	0.773

Comparing the two models, taking into account the parameters of quality, the mediation model is better than the direct model: the SRMR (standardised root mean residual) is improved. The direct model obtains an SRMR of 0.075 and the mediation model obtains an SRMR of 0.061; in both cases, under the threshold established by Henseler et al. [26]. This would imply additional support for the mediating role of absorptive ca-pacity.

5   Conclusions

This work highlights the importance of absorptive capacity in the performance of family businesses. The direct model proposed reveals that the absorptive capacity can explain 36.1% of the performance variance of family businesses. This result is consistent with previous work carried out both in the family business [21, 30] and in non-family businesses [2, 5, 20].

This paper shows the mediating role of innovative capacity in the influence of absorptive capacity on the performance of family businesses. The inclusion of innovative capacity implies strengthening the role of absorptive capacity on the performance of family businesses, increasing the explained variance of the performance of family businesses with the mediation model at 40.6%. This mediating effect of innovative capacity has been demonstrated in previous works on family businesses [31] and in non-family businesses [20]. In addition, this paper shows that absorptive capacity is an antecedent of innovative capacity, the first explaining 50% of the latter. On the other hand, the relevance of the research carried out is that the mediation model completely eliminates the direct effect of the absorptive capacity on the performance of family businesses, which becomes insignificant.

This work presents two pertinent limitations. Firstly, the questions were answered by a single respondent. To overcome this limitation, in this work followed the recommendations of Rong and Wilkinson [32], Woodside [33] and Woodside et al. [34].

The second limitation arises from the sample of businesses used, as they were businesses associated with the Family Business Institute.

As future lines of research, it is proposed to carry out longitudinal studies to analyze the effect of time, or the presence of relatives in the management of the business, or the generational level. It is also proposed to undertake comparative studies with another kind of business and/or country, to check whether there are significant differences.

References

1. Cohen, W.M., Levinthal, D.A.: Absorptive capacity: a new perspective on learning and innovation. Adm. Sci. Q. **2**(5), 99–110 (1990)
2. Jansen, J.J.P., Van den Bosch, F.A.J., Volberda, H.W.: Managing potential and realized absorptive capacity: how do organizational antecedents matter? Acad. Manag. J. **48**(6), 999–1015 (2005)
3. Van den Bosch, F.A.J., Volberda, H.W., De Boer, M.: Coevolution of firm absorptive capacity and knowledge environment: organizational forms and combinative capabilities. Organ. Sci. **10**(5), 551–568 (1999)
4. Floyd, S.W., Lane, P.J.: Strategizing throughout the organization: managing role conflict in strategic renewal. Acad. Manag. Rev. **25**(1), 154–177 (2000)
5. Zahra, S.A., George, G.: Absorptive capacity: a review, reconceptualization and extension. Acad. Manag. Rev. **27**(2), 185–203 (2002)
6. Corona, J., Del Sol, I.: La Empresa Familiar en España (2015), Instituto de la Empresa Familiar, Barcelona (2016)
7. Astrachan, J.H., Shanker, M.C.: Family businesses contribution to the US economy: a closer look. Fam. Bus. Rev. **16**(3), 211–219 (2003)
8. Ringle, C.M., Wende, S., Becker, J.M.: Smart PLS 3. Boenningstedt: SmartPLS GmbH (2015). http://www.smartpls.com
9. Prahalad C.K., Hamel G.: The core competence of the corporation. Harv. Bus. Rev. **68**(3), 79–91 (1990)
10. Teece, D.J., Pisano, G., Shuen, A.: Dynamic capabilities and strategic management. Strateg. Manag. J. **18**(7), 509–533 (1997)
11. Makadok, R.: Toward a synthesis of the resource-based and dynamic-capability views of rent creation. Strateg. Manag. J. **22**(5), 387–401 (2001)
12. Zollo, M., Winter, S.G.: Deliberate learning and the evolution of dynamic capabilities. Organ. Sci. **13**(3), 339–351 (2002)
13. Helfat, C.E., Finkelstein, S., Mitchell, W., Peteraf, M., Singh, H., Teece, D., Winter, S.G.: Dynamic Capabilities: Understanding Strategic Change in Organizations. Wiley, Oxford (2007)
14. Wang, C.L., Ahmed, P.K.: Dynamic capabilities: a review and research agenda. Int. J. Manag. Rev. **9**(1), 31–51 (2007)
15. Lane, P.J., Koka, B.R., Pathak, S.: The reification of absorptive capacity: a critical review and rejuvenation of the construct. Acad. Manag. Rev. **31**(4), 833–863 (2007)
16. Flatten, T.C., Engelen, A., Zahra, S.A., Brettel, M.: A measure of absorptive capacity: scale development and validation. Eur. Manag. J. **29**(2), 98–116 (2011)
17. Prajogo, D.I., Sohal, A.S.: The integration of TQM and technology/R&D management in determining quality and innovation performance. Omega **34**(3), 296–312 (2006)

18. Wiklund, J., Shepherd, D.: Entrepreneurial orientation and small business performance: a configurational approach. J. Bus. Ventur. **20**(1), 71–91 (2005)
19. Chirico, F., Sirmon, D.G., Sciascia, S., Mazzola, P.: Resource orchestration in family firms: investigating how entrepreneurial orientation, generational involvement, and participative strategy affect performance. Strateg. Entrep. J. **5**(4), 307–326 (2011)
20. Yáñez-Araque, B., Hernández-Perlines, F., Moreno-García, J.: From training to organizational behavior: a mediation model through absorptive and innovative capacities. Front. Psychol. **8**, 1532 (2017)
21. Hernández-Perlines, F., Moreno-García, J., Yáñez-Araque, B.: Using fuzzy-set qualitative comparative analysis to develop an absorptive capacity-based view of training. J. Bus. Res. **69**(4), 1510–1515 (2016)
22. Chrisman, J.J., Chua, J.H., Sharma, P.: Trends and directions in the development of a strategic management theory of the family firm. Entrep. Theor. Pract. **29**(5), 555–576 (2005)
23. Barclay, D., Higgins, C., Thompson, R.: The partial least squares (PLS) approach to causal modeling: personal computer adoption and use as an illustration. Technol. Stud. **2**(2), 285–309 (1995)
24. Chin, W.W.: The partial least squares approach to structural equation modeling. Modern Methods Bus. Res. **295**(2), 295–336 (1998)
25. Fornell, C., Larcker, D.F.: Evaluating structural equation models with unobservable variables and measurement error. J. Mark. Res. **18**(1), 39–50 (1981)
26. Henseler, J., Ringle, C.M., Sarstedt, M.: A new criterion for assessing discriminant validity in variance-based structural equation modeling. J. Acad. Mark. Sci. **43**(1), 115–135 (2015)
27. Hair, J.F., Hult, G.T.M., Ringle, C.M., Sarstedt, M.: A Primer on Partial Least Squares Structural Equation Modeling (PLS-SEM). Sage, Thousand Oaks (2017)
28. Preacher, K.J., Hayes, A.F.: Asymptotic and resampling strategies for assessing and comparing indirect effects in multiple mediator models. Behav. Res. Methods **40**(3), 879–891 (2008)
29. Cepeda, G., Henseler, J., Ringle, C., Roldán, J.L.: Prediction-oriented modeling in business research by means of partial least squares path modeling. J. Bus. Res. **69**(10), 4545–4551 (2016)
30. Charão Ferreira, G., Matos Ferreira, J.: Absorptive capacity: an analysis in the context of brazilian family firms. RAM. Rev. de Adm. Mackenzie **18**(1), 174–204 (2017)
31. Meroño-Cerdán, A.L., López-Nicolás, C., Molina-Castillo, F.J.: Risk aversion, innovation and performance in family firms. Econ. Innov. New Technol. **27**(2), 189–203 (2018)
32. Rong, B., Wilkinson, I.F.: What do managers' survey responses mean and what affects them? The case of market orientation and firm performance. Australas. Mark. J. **19**(3), 137–147 (2011)
33. Woodside, A.G.: Moving beyond multiple regression analysis to algorithms: calling for adoption of a paradigm shift from symmetric to asymmetric thinking in data analysis and crafting theory. J. Bus. Res. **66**(4), 463–472 (2013)
34. Woodside, A.G., Prentice, C., Larsen, A.: Revisiting problem gamblers' harsh gaze on casino services: applying complexity theory to identify exceptional customers. Psychol. Mark. **32**(1), 65–77 (2015)

The Mathematics of Interdependence for Superordinate Decision-Making with Teams

W. F. Lawless[(✉)] [iD]

Paine College, Augusta, GA 30901, USA
w.lawless@icloud.edu

Abstract. As a work-in-progress, the *purpose* of this manuscript is to review the function of decision-making as a human process in the field affected by interdependence (additive and destructive social interference). The *scope* of this review is first, to define and describe why interdependence is difficult to grasp intellectually, but much easier intuitively in social contexts (bistability, convergence to incompleteness, non-factorable social states); second, to describe our research accomplishments and applications to hybrid teams (arbitrary combinations of humans, machines and robots); and third, to advance our research by beginning to incorporate the value of intelligence for teams as they strive to achieve a team's superordinate goals (e.g., in the tradeoffs between a team's structure and its effort to achieve its mission with maximum entropy production, MEP). We discuss prior results, future research plans and draw conclusions for the development of theory.

Keywords: Interdependence · Mathematics · Intelligence
Superordinate goals

1 Introduction

1.1 Interdependence

Kahneman [22] concluded that intuition (for rational choice) may often be correct (p. 483), but that it suffers from biases, errors and illusions, and handles uncertainty and competing alternatives poorly (p. 453). In quantum states of superposition or entanglement, intuition fails [15]. Similarly, we argue, intuition is insufficient to rationalize interdependence.

Interdependence is mutual information. It creates bistable, two-sided or many-sided stories. Focusing exclusively on one interpretation of social reality no matter the justification produces an incomplete interpretation; i.e., converged interpretations of social reality create incompleteness. By ignoring alternatives, incompleteness increases uncertainty in decision-making. It is risky to ignore a political opponent's message, a new opponent in a market or a new technology.

© Springer Nature Switzerland AG 2019
E. Bucciarelli et al. (Eds.): DCAI 2018, AISC 805, pp. 91–102, 2019.
https://doi.org/10.1007/978-3-319-99698-1_11

1.2   Prediction

The value of a science is its ability to predict, but the social sciences have long favored statistical explanation over prediction, a trend that big data and machine learning are reversing [20]. However, from Schweitzer [37], while big data and machine learning are able to predict

> what customers will order online … [neither can] be calibrated and validated against real data. … [and both] lack the ability to model the underlying generative mechanisms.

Economic predictions are similarly affected (e.g., [31]). Per Sen [36], economic "predictions are notoriously unreliable." Based on the inability of economists to construct a successfully predictive science, from Appelbaum [1], "a growing number of experts, including some Federal Reserve officials, say it is time for the Fed to consider a new approach to managing the economy." From Samuelson [35], "The record of economists, including those at the Federal Reserve, over the past half century has been discouraging." Sharma [38] captures this problem for economists in the title of his article: "When Forecasters Get It Wrong: Always". Andy Haldane, chief economist for the Bank of England [25], stated that

> "Economic forecasts on the eve of the credit crunch and the Great Recession were … "not just wrong but spectacularly so."

Some might argue that game theory predictions are at least reliable; e.g., regarding competition, "the pursuit of self-interest by each leads to a poor outcome for all" ([2], pp. 7–8). Competition can be avoided, Axelrod continued, when sufficient punishment exists to discourage it. There are two problems with Axelrod's conclusion: the notion that punishment is necessary to gain cooperation is reprehensible; and while reliable, it is not valid: the "evidence of mechanisms for the evolution of cooperation in laboratory experiments … [is not found in] real-world field settings" ([34], p. 422).

Why is the inability of social scientists and economists to make valid predictions about social reality a problem? Hybrid teams are coming (arbitrary combinations of humans, machines and robots). Yet, for hybrid teams to work effectively, mathematical models are essential. It may be possible with machine learning to have swarms of robots to attack an enemy, as happened for the first time against a Russian base in Syria [18]. But these teams are inefficient; cannot scale, not innovative nor evolving.

We are developing an alternative theory of the interaction around a model of interdependence and competition with real data that produces accurate predictions [27–29]. We have found that predicting outcomes based on methodological individualism, the prevailing or traditional model, is non-predictive; in contrast, we propose that based on the structure of teams, organizations or systems (nation states), outcomes are more predictable.

1.3   Competition and Information

Like Axelrod, Krugman [24] has argued that competition among nations might lead to trade wars and xenophobia. But then where does the information for successful prediction come? Instead, we see competition as the source of information arising from an idea, team, organization or system withstanding attacks in a marketplace of competing forces. Justice Ginsburg [16] argued that the Supreme Court should not take a case until it had been tested by an "informed assessment of competing interests" (p. 3). Scientists commonly use game theory to test ideas regarding competition, but game theory while bistable between individuals has been discredited as being complete, lacking uncertainty and unable to scale to teams and organizations [27]. Moreover, games ignore the struggle for survival; surviving in business, science, the law and other fields is always a struggle (e.g., the health market; in [14]).

Rejecting the value of meaning, Shannon's [39] information theory determined how signals could be transmitted (from interdependence or mutual information) through a channel with the maximum amount of information that could be sent. Shannon information requires independent, identical data (*i.i.d.*) and agents; i.e., correlations should not exist among *iid* agents, and if they do, these correlations should be removed (e.g., detrending a time series). For our purposes, Shannon concluded that signals could be compressed by reducing redundant information; but that some redundancy was necessary for accurate messages. Based on Shannon, the information generated for the signals transmitted magically appear rationally; but since agents are *iid*, and messages only traverse channels, how can opponents ever learn anything? In our field theory, competitive actions generate the information naturally that is mutually transmitted to players and observers alike, a cost minimized by team boundaries and information orthogonal to observation. Whereas the information transmitted between any two *i.i.d.* agents in a team remains valid under Shannon communication, however, even the best teams experience conflict [17]; and a team often has roles orthogonal to others in a team (e.g., a football quarterback, center, tackle), preventing the information from being transmitted when generated by an orthogonal role, a bound to rationality.

Other bounds exist. Per Simon [40], expertise is bounded by knowledge: no knowledge, no expertise ([40], pp. 6–7); bounded rationality places a limit upon the ability of human beings to adapt optimally, or even satisfactorily, to complex environments ([40], p. 12). But Simon does not distinguish between skills-based knowledge and the knowledge articulated like an algorithm. If these two knowledges are orthogonal, a path forward opens.

The independence of agents is easily described with Shannon [39] but, except for slaves [9], the state of a team of independent (non-correlated) agents becomes harder to fathom (e.g., holonic).[1] In Shannon, the joint entropy is never less than the state of any one agent (Eq. 1)

$$H_{A,B} \geq H_A, H_B \tag{1}$$

But with Von Neumann entropy, we can fathom real team effects, like constructive and destructive interference, allowing a team to be described as interdependent, meaning that the elements of a team cannot be factored but are correlated, and

---

[1] Simultaneously, a holon is a whole unto itself and a part of a larger group; it is a self-contained entity and an element in a larger system.

importantly that the entropy generated by one agent can offset that of another, known as subadditivity (Eq. 2).

$$S_{A,B} \leq S_A + S_B \tag{2}$$

In contrast, if the teammates of a team or organization have orthogonal roles, correlations go to zero (Eq. 3). But the role information in the least redundant team is orthogonal (e.g., conductor, first violinist, pianist), the information shared by the best teammates becomes subadditive, producing a poor correlation:

$$lim_{N \to \infty} r_{teammates} \to 0 \tag{3}$$

If action and observation (beliefs) are interdependent, Eq. 3 accounts for the poor correlations found between action and beliefs about action [45]. But given a super-ordinate goal which is an overarching goal that subsumes subgoals, the subgoals aim to satisfy the superordinate goal. For a superordinate goal, the lack of a correlation at the team level flips for a superordinate goal,

$$lim_{N \to \infty} r_{SuperordinateGoal} \to 1 \tag{4}$$

A superordinate goal reintroduces a strong correlation between the cognition of teammates and their mission (Eq. 4), guiding them to produce MEP based on information about shared superordinate goals. If these findings stand, Eqs. 3 and 4 underwrite the metrics for hybrid teams. Superordinate goals underscore the value of intelligence for constructing and operating teams (e.g., [42]).

1.4   Beliefs and Structures. Superordinates

To achieve maximum entropy production (MEP), superordinate beliefs support the structures wherein these beliefs operate. When structures fail, superordinate beliefs are impaired; e.g., [4],

> Should there be a change of control of Xerox, it would mark the end of the independence of a stalwart of American industry that was an early technological trailblazer that has been bedeviled by a drop off in demand ...

What has happened at Xerox has happened economically with the Palestinians [7]:

> [The] Palestinian Authority's (PA) ... for the foreseeable future will continue to rely heavily on donor aid for its budgetary needs and economic activity." The failure to develop new technology can also impair a superordinate goal [41]: "Volkswagen AG is near resolving a criminal investigation into the German auto giant's emissions cheating ... that would require the company to pay a financial penalty of several billion dollars and cap a major aspect of a long-running crisis.

1.5   Interdependence Replaces Methodological Individualism

Kang [23] dismissed a new book by stressing the dangers of implicit bias for organizations: "Over the past two decades, science has demonstrated that implicit social cognitions exist, are pervasive, are predictably biased, and alter judgments that produce discriminatory decisions" (see also [21]). But the idea that implicit bias has a scientific basis is without merit [5]. Kang's review steps back into a methodological individualism (MI) that ignores the hard-won structural remedies afforded by the law.

Not surprisingly, Kang's [23] "invisible reach" of implicit bias can only be corrected with well-paid consultants [3]. But fixing a made-up problem is impossible. "There's a growing skepticism about whether unconscious bias training is ... effective ... " [13], but consultants haven't given up hope.

For decades, public funds were wasted by schools and businesses failing to raise each individual's self-esteem [12]. That ended after 30 years when surveys of self-esteem were found to be negligibly correlated with work or academics [3], a finding that Kang and the many well-paid consultants must ignore.

Widespread problems plague the predictions based on the individualism inherent in traditional social science; e.g., political polls fail to predict [6]; experimental social science struggles with the failure to replicate [33]; market predictions fail to impress [31]; and even *Science* has reported on the failure of medical interviews contradicted by compliance behaviors [8].

As part of a news report in *Science*, Cohen [8], reported that in an HIV prevention drug trial, the women partners reported complying with the drug regimen 90% of the time, indicating that the drug had failed to protect the women from contracting AIDS from their partners. But when the researchers later analyzed blood levels of drugs in the women, they found that no more than 30% had evidence of anti-HIV drugs in their body at the time of their self-reports. "There was a profound discordance between what [the women had] told us ... and what we measured," infectious disease specialist Jeanne Marrazzo said.

The problem is not implicit bias, but the natural interference arising from the interdependence between self-reports and behavior, a problem with all questionnaires (e.g., [45]). Managing the interference from interdependence is key to team performance, whether in science, politics or the law [27]; to promote his argument, Kang had to demonize legal structural remedies, but they not only work, structures are predictive, although their size is an open intellectual problem [10], one that we have solved mathematically [28]; viz., given that the best teams are highly interdependent [11], which maximizes Shannon-type communication between any two members of a team [39], any team larger than the minimum number for maximum communication will impair communication and efficiency. Unlike implicit bias which is irrelevant, interdependence as a mathematical theory has to be solved for the future of effective and efficient hybrid teams (arbitrary combinations of humans, machines and robots).

1.6   Conflict Between Superordinate Goals Leads to Discord and Disarray

Conflict between superordinate goals leads a team to discord and disarray; e.g., [19]:

A century ago ... thousands of U.S. combat forces found themselves in the middle of Russia's communist revolution, fighting on Russian soil for months in a losing battle against the Bolsheviks. ... an early instance of the perils of small-scale military interventions abroad without a clear mission or support at home. .... more than 200 soldiers died, mostly from combat. By the spring of 1919, months after World War I had ended, conditions on the Arkhangelsk front were difficult, and morale was low. American soldiers began to resist orders.

1.7   Prior Research

With two sides interdependent to every decision, how can decisions best engage stakeholders (citizens, politicians, professionals, scientists, managers); how can action be expeditious; and how important is leadership? In prior research of stakeholders making decisions on how best to recommend that military nuclear wastes be disposed [26], we contrasted the outcomes of recommendations made by the Department of Energy's Hanford Citizens Advisory Board in the State of Washington with those by DOE's Savannah River Site's CAB in South Carolina. Hanford and SRS have the largest cleanup budgets in the USA, including high level radioactive wastes (HLW) from reprocessing spent military nuclear fuel. To illustrate the difference between these two sites, the technology to vitrify HLW was invented at Hanford but used at SRS continuously from 1996 to vitrify over half of its HLW, whereas Hanford has yet to vitrify its first batch of HLW. We concluded that, compared to the majority rules (MR) used by SRS's CAB, the consensus-seeking rules (CR) used by Hanford's CAB recommended fewer concrete decisions that slowed cleanup processes and generated less transparency. Similarly, the European Union concluded that a minority under CR is capable of blocking a majority from making concrete decisions [44].

In recent research on mathematical models of teams and organizations using Kullback-Leibler divergence, we found that (oil firms; in [27]; and militaries; in [28]) organizations operating under autocratic governance have significantly more redundant workers compared to organizations governed under a democracy. Further, we found significantly less competition and more corruption under governing autocracies. Taken together, these results reject the rational model based on methodological individualism (MI), but they are in agreement with the second law of thermodynamics. We believe that these two findings can be integrated: Theoretically, we predict that CR creates more redundant deciders (disorganized superordinate vectors) whereas MR integrates neutrals with local and global superordinate goals by using interdependence to entangle neutrals into supporting decisions that promote worker safety, citizen welfare and a sustainable environment.

Measuring Kullback-Leibler divergence between separate distributions of oil firms and militaries, we found significant support for predicting that redundant teammates proliferate in teams (organizations) governed under autocracies to satisfy constituencies, increasing corruption but reducing efficiency. Redundancy impedes the interdependence necessary for effective and efficient teams. Assuming that Shannon information [39] distributes asymmetrically among divergent roles (e.g., pitcher, catcher, first baseman), separating structure and performance, we discovered that team size (structure) follows the second law of thermodynamics, solving the open problem of team size; e.g., for an effective organizational merger, its transformed structure emits less entropy while entropy is removed to the environment (by reducing employees); but when a firm like General Electric plans to break apart, GE's subsequent parts require more employees and structural entropy. Before breaking apart, GE suffers from redundant elements that reduce interdependence, presaging failure.

To generalize, for maximum performance, the entropy of an optimum team structure decreases, implying that as teammates operate as a unit, in the limit, their degrees of freedom (*dof*) become

$$lim_{N \to \infty} log(dof) \to 0 \qquad (5)$$

Further, since the role of information in the least redundant teams is orthogonal (e.g., conductor, first violinist, pianist), the information shared by the best teammates becomes subadditive, a poor correlation results (Eq. 3), while for the team to be able to produce MEP, information about its superordinate goal must be shared by team members (Eq. 4). If true, these findings will become the metrics for hybrid teams.

2 Method

2.1 Work in Progress. Intelligence and Teamwork

In prior research, we have established the value of an arbitrary demarcation of interdependence to better appreciate its complexity and our advancement. In Table 1, we list its arbitrary factors.

Table 1. An arbitrary demarcation of interdependence [27, 28].

Factors	Examples
Bistability	Stories with two sides; e.g., by prosecutors and defense attorneys; pro-Einsteinian and pro-Bohr quantum interpretations; and conservative and liberal politicians
Incompleteness	As a story constructs its theme around well-illustrated images of a protagonist fighting the odds against a strong adversary, it gains strength as it reaches a climax, but, consequently, increasing uncertainty in the part of the story that does not fit its theme; e.g., a hit movie about a war often shows one side of a story as good, but ignoring the other side except to exploit it as the enemy and to show the morality of the protagonist
Non-factorability	For a troubled marriage (or a business losing market share), before divorcing, a counselor tries to uncover the cause of their trouble with extensive discussions over a long period of time; similarly, a business merger combines objective data with subjective judgments to determine whether a merger target is a good fit that can improve its market competitiveness

We have theorized that non-factorability is reflected as responsiveness. Highly interdependent teams should be the most responsive to the external (e.g., market) and internal (e.g., internal communication) forces that they face; given freedom for the labor and capital available to a firm to address the market forces that it faces, this prediction means that team member redundancy must be at a minimum to maximize the communication among a team's members while minimizing its costs for communication (low internal interference); i.e., the size of an organization is the minimum size necessary to accomplish a team's mission for firms in free economies versus non-free economies, which we found for oil firms [27] and the size of a nation's military [28]. For both oil firms and militaries, we also found that the less free was an economy, the larger the size of the teams and the more corrupt were the societies.

For this new study, we sought to add to our past findings by exploring intelligence with human development and gross domestic product (GDP) to reflect MEP. We studied the UN's Middle East North Africa (MENA) countries based on its Human development report [46].[2] Adding Israel to the MENA data with the UN's Human Development Index (HDI) data and the mean years of schooling for citizens,[3] we found a significant correlation ($r = .90$, $p < .001$, two tailed). Comparing HDI for these 20 countries with GDP from the World Bank[4] we found a significant correlation ($r = .67$, $p < .001$, two tailed); and we found a significant correlation between mean years of schooling and GDP ($r = .62$, $p < .01$, two tailed).

3   Future Research

For future research, if teammates are characterized as intelligent things, whether humans, machines or robots, as a key step in the engineering of teams, we expect to find that larger team structures generate more entropy (i.e., more arrangements are possible), requiring proportionately more energy to cohere; that perfect teams operate at stable, ground states, dysfunctional teams at excited, emotional states; and that entropy for a perfect team is subadditive (i.e., Eq. 2); Von Neumann subadditivity occurs mindfully, we speculate, with superposition of the terms from interference, an information loss from maximum interdependence (precluding the replication of a perfect team, similar to no-cloning; e.g., [43], p. 77), while the information from a dysfunctional team increasingly gains Shannon information to inadvertently clarify context (Eq. 3), the two forming a metric for team performance (Fig. 1).

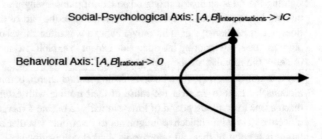

Fig. 1. Based on theory and findings, we place physical reality on the Behavioral axis (x-axis) for agreement on physical behavior. We place the multiple interpretations of behavior on the imaginary axis (y-axis). The stronger are imaginary effects, the more a dispute is oscillating; e.g., the DOE and NRC turf battles described below

---

[2] Algeria, Bahrain, Egypt, Iran, Iraq, Jordan, Kuwait, Lebanon, Libya, Morocco, Palestine, Oman, Qatar, Saudi Arabia, Sudan, Syria, Tunisia, United Arab Emirates, Yemen; from http://hdr.undp.org/sites/default/files/hdrp_2010_26.pdf.

[3] HDI; from http://hdr.undp.org/en/composite/HDI.

[4] World Bank, gdp and populations; https://data.worldbank.org/country.

"Turf battles" arise between nearly equal centers of power (e.g., [30]), displayed as undamped behavioral oscillations similar in character to sine waves or to those undamped predator-prey oscillations modeled in biology.

As a classic example of a turf battle taken from a public meeting held in 2011 [26], the State of South Carolina complained in public that the U.S. Department of Energy (DOE) was unlikely to close two high-level radioactive waste tanks by 2012 as had been legally agreed upon by the three agencies involved (DOE; SC; and U.S. EPA), the slipping milestone was caused by a set of unresolved challenges from the Nuclear Regulatory Commission (NRC) (NRC was injected into the tank closure decision process by the 2005 NDAA.[5] However, once the public got involved in this "turf battle", all agencies including the NRC quickly agreed to DOE's closure plan and the tanks were closed in what one DOE official described as "...the fastest action I have witnessed by DOE-HQ in my many years of service with DOE."

3.1 Discussion and Conclusion

Testing an idea by challenging it with the "informed assessment of competing interests" [16] is the only path to knowledge, where knowledge is signified by producing zero entropy ([39]; also, [9], p. 244). This conclusion is related to Shannon's concept of the maximum compression of information by reducing redundancy [39]; it is related to the use of intelligence in market economies to reduce the size of teams by shrinking redundancy ([27, 28]), and we argue, it is comparatively true for organizations, businesses, innovations or systems like nations. It is also indirectly related to Markowitz's [32] portfolio theory, the object being for investors to reduce their risk in the market by spreading the risk around by investing in multiple investments with low correlations among them.

What can be measured mathematically about the most competitive organizations and generalized to hybrid teams (i.e., arbitrary combinations of humans, machines and robots)? All things equal, compared to democracies, we conclude that organizations governed by autocracies are quieter, less able to innovate and produce less maximum entropy (MEP), reducing social welfare, a conundrum for rational designs of hybrids.

References

1. Appelbaum, B.: As Economy Strengthens, Fed Ponders New Approach, New York Times, 9 January 2018. https://www.nytimes.com/2018/01/09/us/politics/federal-reserve-inflation.html
2. Axelrod, R.: The evolution of cooperation. Basic, New York (1984)

---

[5] Viz., the National Defense Authorization Act, SEC. 3116. DEFENSE SITE ACCELERATION COMPLETION: "the Secretary of Energy ... in consultation with the Nuclear Regulatory Commission" put the NRC on an equal footing with the DOE in matters of closing "high-level radioactive waste" tanks. From https://www.gpo.gov/fdsys/pkg/PLAW-108publ375/pdf/PLAW-108publ375.pdf.

3. Baumeister, R.F., Campbell, J.D., Krueger, J.I., Vohs, K.D.: Exploding the self-esteem myth. Sci. Am. **292**(1), 84–91 (2005). https://www.uvm.edu/~wgibson/PDF/Self-Esteem%20Myth.pdf
4. Benoit, D., Cimilluca, D., Mattioli, D.: Xerox Is in Talks for a Deal With Japan's Fujifilm. Two companies discussing array of possible deals that may or may not include a change of control of Xerox. Wall Street J. 2018, 1/10. https://www.wsj.com/articles/xerox-is-in-talks-for-a-deal-with-japans-fujifilm-1515631019
5. Blanton, H., Klick, J., Mitchell, G., Jaccard, J., Mellers, B., Tetlock, P.E.: Strong claims and weak evidence: reassessing the predictive validity of the IAT. J. Appl. Psychol. **94**(3), 567–582 (2009)
6. Byers, D.: Nate Silver: Polls are failing us. Politico, 8 May 2015. http://www.politico.com/blogs/media/2015/05/nate-silver-polls-are-failing-us-206799.html
7. CIA: World Fact Book: West Bank, Central Intelligence Agency (2015). https://www.cia.gov/library/publications/the-world-factbook/geos/we.html
8. Cohen, J.: Human Nature Sinks HIV Prevention Trial. Science **351,** 1160 (2013). http://www.sciencemag.org/news/2013/03/human-nature-sinks-hiv-prevention-trial
9. Conant, R.C.: Laws of information which govern systems. IEEE Trans. Syst. Man Cybern. **6**, 240–255 (1976)
10. Cooke, N.J., Hilton, M.L. (eds.) Enhancing the Effectiveness of Team Science. Authors: Committee on the Science of Team Science; Board on Behavioral, Cognitive, and Sensory Sciences; National Research Council. National Academies Press, Washington, DC (2015)
11. Cummings, J.: Team Science Successes and Challenges. National Science Foundation Sponsored Workshop on Fundamentals of Team Science and the Science of Team Science (June 2), Bethesda MD (2015). https://www.ohsu.edu/xd/education/schools/school-of-medicine/departments/clinical-departments/radiation-medicine/upload/12-_cummings_talk.pdf
12. Diener, E.: Subjective well-being. Psychol. Bull. **95**(3), 542–575 (1984)
13. Emerson, J.: Don't Give Up on Unconscious Bias Training – Make It Better. Harvard Business Review, 28 Apr 2017. https://hbr.org/2017/04/dont-give-up-on-unconscious-bias-training-make-it-better
14. Farr, C.: Castlight Health CEO warns Jeff Bezos that health tech is a 'tough business'. CNBC Tech (2018, 2/4). https://www.cnbc.com/2018/02/04/castlight-health-ceo-bezos-will-learn-health-care-is-a-tough-business.html
15. Gershenfeld, N.: The Physics of Information Technology. Cambridge University Press (2000)
16. Ginsburg, R.B.: American Electric Power Co., Inc., et al. v. Connecticut ET AL., 10-174 (2011). http://www.supremecourt.gov/opinions/10pdf/10-174.pdf
17. Hackman, J.R.: Six common misperceptions about teamwork. Harvard Business Review (2011). https://hbr.org/2011/06/six-common-misperceptions-abou
18. Hambling, D.: A swarm of home-made drones has bombed a Russian airbase. New Scientist (2011). https://www.newscientist.com/article/swarm-home-made-drones-bombed-russian-airbase/?cmpid=NLC%7CNSNS%7C2018-0111-GLOBAL&utm_medium=NLC&utm_source=NSNS
19. Hodge, N.: When U.S. Troops Battled Bolsheviks. An early example of mission creep, the incursion still irks Russians a century later. Wall Street Journal (2018, 1/26). https://www.wsj.com/articles/when-u-s-troops-battled-bolsheviks-1516980334
20. Hofman, J.M., Sharma, A., Watts, D.J.: Prediction and explanation in social systems. Science **355**, 486–488 (2017)

21. Huet, E.: Rise of the Bias Busters: How Unconscious. Bias Became Silicon Valley's Newest Target. Forbes 2 November 2015. https://www.forbes.com/sites/ellenhuet/2015/11/02/rise-of-the-bias-busters-how-unconscious-bias-became-silicon-valleys-newest-target/#7c41ec5d19b5

22. Kahneman, D.: Maos of bounded rationality: A perspective on intuitive judgment and choice. Prize Lecture (2002, 12/8). http://www.nobelprize.org/nobel_prizes/economic-sciences/laureates/2002/kahnemann-lecture.pdf

23. Kang, J.: Book Review. Kahn, J. Race on the brain. What implicit bias gets wrong about the struggle for racial justice. Science **358**(6367), 1137–1138 (2017)

24. Krugman, P.: Competitiveness: a dangerous obsession. the hypothesis is wrong. Foreign Affairs (1994). https://www.foreignaffairs.com/articles/1994-03-01/competitiveness-dangerous-obsession

25. Lanchester, J.: The Major Blind Spots in Macroeconomics, New York Times Magazine, 7 February 2017. https://www.nytimes.com/2017/02/07/magazine/the-major-blind-spots-in-macroeconomics.html?

26. Lawless, W.F., Akiyoshi, M., Angjellari-Dajcic, F., Whitton, J.: Public consent for the geologic disposal of highly radioactive wastes and spent nuclear fuel. Int. J. Environ. Stud. **71**(1), 41–62 (2014)

27. Lawless, W.F.: The entangled nature of interdependence, Bistability, irreproducibility and uncertainty. J. Math. Psychol. **78**, 51–64 (2017)

28. Lawless, W.F.: The physics of teams: Interdependence, measurable entropy and computational emotion. Front. Phys. **5**, 30 (2017). https://doi.org/10.3389/fphy.2017.00030

29. Lawless, W.F.: Interdependence, decision-making, superordinate vectors and sustainability. A comparison of SRS and Hanford military nuclear waste sites. Invited talk: Sustainable and Renewable Energy Research 2018, Paris, France, Euroscicon, August 2018

30. Light, D.W.: Turf battles and the theory of professional dominance. Res. Sociol. Health Care **7**, 203–225 (1988)

31. Mackintosh, J.: Wall Street's 2017 Market Predictions: Pathetically Wrong. Forecasting is difficult, but this year showed exactly how pointless it can be: Markets performed opposite of virtually all predictions. Wall Street J., 23 Nov 2017 https://www.wsj.com/articles/wall-streets-2017-market-predictions-pathetically-wrong-1511474337

32. Markowitz, H.M.: Portfolio selection. J. Financ. **7**(1), 77–91 (1952)

33. Nosek, B., corresponding author from OCS, Open Collaboration of Science: Estimating the reproducibility of psychological science. Science **349**(6251), 943 (2015). supplementary: 4716-1 to 4716-9 (2015)

34. Rand, D.G., Nowak, M.A.: Human cooperation. Cogn. Sci. **17**(8), 413–425 (2013)

35. Samuelson, R.: Does the Federal Reserve Need a New Playbook?. Washington Post, 15 Feb 2018. https://www.realclearmarkets.com/articles/2018/01/15/does_the_federal_reserve_need_a_new_playbook_103101.html

36. Sen, A.K.: Prediction and economic theory. Proc. Roy. Soc. Lond. A **407**, 3–23 (1986)

37. Schweitzer, F.: Sociophysics. To the extent that individuals interact with each other in prescribed ways, their collective social behavior can be modeled and analyzed. Physics Today 71, 2, 40 (2018). https://doi.org/10.1063/PT.3.3845

38. Sharma, R.: When Forecasters Get It Wrong: Always. New York Times (2017). https://www.nytimes.com/2017/12/30/opinion/sunday/when-forecasters-get-it-wrong-always.html

39. Shannon, C.E.: A mathematical theory of communication. Bell Syst. Tech. J. **27**(379–423), 623–656 (1948)

40. Simon, H.A.: 9/23), Bounded rationality and organizational learning, Technical Report AIP 107. CMU, Pittsburgh, PA (1989)

41. Viswanatha, A., Spector, M.: Volkswagen Near Settling U.S. Criminal Case Over Emissions Cheating. Settlement with Justice Department could come as soon as next week. Wall Street J., 6 Feb 2017. http://www.wsj.com/articles/volkswagen-near-settling-u-s-criminal-case-over-emissions-cheating-1483724523

42. Wissner-Gross, A.D., Freer, C.E.: Causal entropic forces. Phys. Rev. Lett. **110**(168702), 1–5 (2013)

43. Wooters, W.K., Zurek, W.H.: The no-cloning theorem, Physics Today, pp. 76–77, February 2009. http://www.physics.umd.edu/studinfo/courses/Phys402/AnlageSpring09/TheNoCloningTheoremWoottersPhysicsTodayFeb2009p76.pdf

44. WP: White Paper. European governance (COM (2001) 428 final; Brussels, 25.7.2001). Brussels, Commission of the European Community (2001)

45. Zell, E., Krizan, Z.: Do people have insight into their abilities? a metasynthesis? Perspect. Psychol. Sci. **9**(2), 111–125 (2014)

46. Salehi-Isfahani, D.: Human Development in the Middle East and North Africa, Human Development Research paper, Table 1, p 36, 20 Oct 2016. http://hdr.undp.org/sites/default/files/hdrp_2010_26.pdf

Towards a Natural Experiment Leveraging Big Data to Analyse and Predict Users' Behavioural Patterns Within an Online Consumption Setting

Raffaele Dell'Aversana[1(✉)] and Edgardo Bucciarelli[2]

[1] Research Centre for Evaluation and Socio-Economic Development, University of Chieti-Pescara, Viale Pindaro 42, 65127 Pescara, Italy
r.dellaversana@gmail.com
[2] Department PPEQS – Section of Economics and Quantitative Methods, University of Chieti-Pescara, Viale Pindaro 42, 65127 Pescara, Italy
edgardo.bucciarelli@unich.it

Abstract. The authors develop a model for multi-criteria evaluation of big data within organisations concerned with the impact of an ad exposure on online consumption behaviour. The model has been structured to help organisations make decisions in order to improve the business knowledge and understanding on big data and, specifically, heterogeneous big data. The model accommodates a multilevel structure of data with a modular system that can be used both to automatically analyse data and to produce helpful insights for decision-making. This modular system and its modules, indeed, implement artificial intelligent algorithms such as neural networks and genetic algorithms. To develop the model, therefore, a prototype has been built as proof-of-concept using a marketing automation software that collects data from several sources (public social and editorial media content) and stores them into a large database so as the data can be analysed and used to implement business model innovations. In this regard, the authors are conducting a natural experiment - which has yet to be completed - to show that the model can provide useful insights as well as hints to help decision-makers take further account of the most 'satisficing' decisions among alternative courses of action.

Keywords: Computational behavioural economics
Online consumption setting · Natural experiments in economics
Big data · Computational intelligence

JEL codes: C81 · C99 · D12 · D22

1 Introduction

Although it is currently being experimented on real cases and opened to further experimentations in microeconomics, this paper is conceived in the framework of the theory of computation and computational complexity which identifies, largely, the foundations of computational behavioural economics (for an insightful survey, see [1])

© Springer Nature Switzerland AG 2019
E. Bucciarelli et al. (Eds.): DCAI 2018, AISC 805, pp. 103–113, 2019.
https://doi.org/10.1007/978-3-319-99698-1_12

and classical behavioural economics (for a critical discussion, see [2]). In this framework and based on previous studies, *e.g.* [3–5], the authors build a model including measurable indicators with the aim of designing and running a natural experiment, and analysing its outcome. The main research goal is to encompass a reproducible model to be used effectively in order to manage and analyse big data together with the evolution of their economic value over time. The model is made up of several integrated parts: (i) definition of the theoretical framework; (ii) data collection; (iii) data elaboration and analysis; (iv) reporting; (v) knowledge database and improvement actions (implementation and follow up).

More specifically, the authors integrate their model in a marketing automation software (MA). As is common knowledge, MA helps organisations stay connected with their audience automatically. In particular, it refers to technologies designed for marketing departments to more effectively market on multiple channels and automate repetitive tasks. In this business area, organisations regularly use electronic communications and digital information processing technology in business transactions (e-commerce) to define and redefine relationships for value creation both between or within organisations, and between organisations and final users. In doing so, organisations promote their activities through social networking and microblogging services, trying both to increase the number of users and to convert them into followers and consumers or, better yet, into customers. In addition to publishing contents on social networks and blogs, organisations customarily plan and implement direct marketing policies on proprietary and licensed databases, by which promoting their products-or-services, even focusing on specific targets, considering indistinctly all potential users or performing a stratified analysis or a meta-analysis of them. The MA automatically collects large dataset from the market (through e-commerce and social channels) including subjective information on final users, be they anonymous, simple followers, sporadic consumers, or loyal customers. Moreover, the MA provides advanced tools for data analysis. These tools enable social networking and microblogging management (*e.g.*, publishing and editing) as well as marketing automation services (*e.g.*, lead scoring, automated drip campaigns, and CRM integration). In this paper, as mentioned earlier, the authors show an integration between the MA and the model proposed by them. Through this integration, all the data collected by the MA will flow into the model for subsequent data-mining. In particular, the model provides automated analysis tools via artificial intelligent algorithms (especially based on deep learning algorithms). Through these tools the model will automatically analyse and predict users' behavioural patterns, suggesting the best marketing actions for each of them in an online microeconomic setting.

2   Data Structure

To describe how the model works and how it integrates with the MA, we need to introduce, briefly, the structure of the MA database and the differences with the model database proposed, formally designed in Sect. 4. To get to the point and ease the understanding, we start from a typical use case, where the MA database is equipped with the data of a hypothetical organisation having (i) an e-commerce website;

(ii) a blog or microblog website where news of interest is published for some potential final users; (iii) a few social network channels; and (iv) a mailing list to send newsletters. For each user, the MA database collects all the actions she makes. For instance, for a specific user, we have knowledge about any comments posted, *likes* added, newsletters opened, links clicked on the e-commerce platform and financing services, products/services bought on it, and so on. For each action performed by the user, furthermore, the date, time as well as usage time are shown so as to sort each action in various orders (*e.g.*, chronological order). Following the chronological order, researchers can discover, among others, that a single user might have behaved anonymously on the social networks for a certain period of time, then she might have subscribed to the newsletter service on the blog, might have acquired a certain content of interest (*e.g.*, on certain newsletters), and finally might have bought some products/services. In the meanwhile, she might have also surfed the e-commerce platform looking for specific products and/or services and visiting specific web pages. For this reason, there are a lot of data available for each user registered into the database, thus actually researchers have to deal with a huge amount of data, namely, big data. This applies to all types of organisation, not just those oriented to the sale of products/services, in order to enhance their mission readiness and familiarisation of those forces operating in a certain microeconomic and social environment.

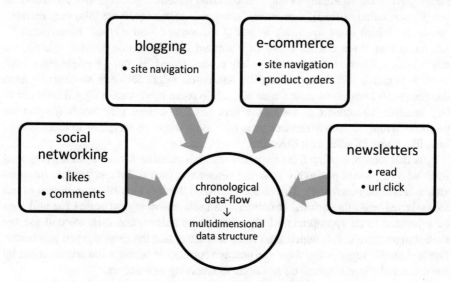

Fig. 1. The MA database is built by continuous automatic collection of big data from multiple sources within an online consumption setting. The multidimensional structure, discussed in Sect. 4, is built from the MA database as a single chronological data-flow.

All these data are transformed and imported as a single chronological data-flow in a large database, structured as a multidimensional data structure that can be viewed as a tensor in the way detailed in Sect. 4. The data-flow structured as multidimensional data gives us the opportunity to study users' behavioural pattern, and to predict what will be the most 'satisficing' decisions [6] among alternative course of action and, thus, the most 'satisficing' marketing actions to sort out by the organisation. If the database is big enough, that is, if it has many registered users with a lot of data for each of them, it becomes possible to better analyse their behavioural pattern, discover possible strategies, and use these patterns to predict possible targets and suggest further marketing actions.

3 Automated Data Analysis

With the aim to move beyond the manual analysis of final users' behavioural data mostly performed by humans, and thus implement automated analysis, note that in the MA the researcher can analyse data basically doing manual mining, looking at pre-defined reports, and then deciding what might be the most 'satisficing' decisions among alternative course of marketing action (the final objective is usually to increase sales). However, many organisations, even small to midsize businesses, tend to have multiple lines of ongoing business, each with its own marketing operations. In the course of data collection and analysis processes, therefore, what we are working on is to start with a specific research question applied to microeconomics of organisations (think, for example, of the causal inference in microeconomics and marketing [4, 5, 7]) and try to figure out what interesting themes are automated analysis of data using typical deep learning AI tools, especially neural networks and genetic algorithms (for an overview, see [8], particularly [9–13]). For example, suppose to answer to the following research question: *"Which users are likely to buy if we contact them through a newsletter?"*. We know that if we contact every user without applying some filtering criteria, we might intercept some users that might buy products/services, but we might even incur possible negative effects on profitability: some users might decide to unsubscribe from the newsletter because of their frequency, while some other users might decide not to buy because the content of the newsletters might not meet their needs (maybe we promote 'wrong' products/services or maybe users might be receptive to incentives to buy, like a special offer or a discount).

For that reason, starting from a specific research question we aim at identifying final users who responded positively to earlier newsletters (that is, bought products/services after receiving a newsletter). This is accomplished through data filtering criteria in our tensor-based multidimensional structure. The users chronological action list will then be submitted to an appropriate AI algorithm to be trained, and then we will use the trained algorithm to find which final users to contact and the most tailored newsletter. Our first results suggest that there is a positive correlation between the actions made by the users and several optimistic reactions to receiving newsletters.

In a nutshell, the algorithm and the related integration schema can be represented as follow. After identifying a specific research question, we find out – by standard mining on the multidimensional data structure – the final users within the database that satisfied similar research questions in the past. The big database is split in two datasets: the first is used as input to develop appropriate AI algorithms, while the second is used to verify the effectiveness of trained algorithms. Once we find an effective algorithm, this can be integrated into the MA in order to have a type of decision support system so that the algorithm can automatically suggest marketing actions without further manual intervention made by the researchers (Fig. 2).

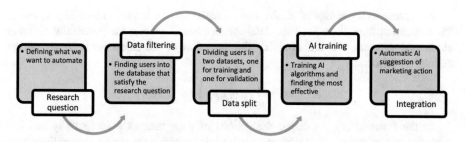

Fig. 2. The process to implement the AI automation into the MA software.

As in a previous framework [3], we define multilevel indicators to help find an answer to the research question. However, these indicators are built not only starting from measurable quantities but they could come from the output of AI algorithms, too. To ease the integration of our model into the MA software, we decided to let the model work as a black box: the algorithm that answers a specific research question can be written in any language (we are working, *e.g.*, with Haskell and Scala programming languages [14, 15]). The only requirement is to adhere to an application programming interface (API) according to a specification that enables the ability to communicate with our model, integrated into the MA.

4 Design of the Model Database

Designing more formally the architecture of our large database and how it is structured by our model data collection, we denote the representation of our database by D, whereby, we have:

$$D = \{C, A\}$$

$$C = \{C_1 \ldots C_i \ldots C_n\}$$

$$A = \{A_1 \ldots A_i \ldots A_n\}$$

where $C = \{C_1 \ldots C_i \ldots C_n\}$ is the set of final users, $n = |C|$ is the cardinality of that set, and $A = \{A_1 \ldots A_i \ldots A_n\}$ is the tensor of all the actions registered for each user (see Fig. 1). In other words, each user C_i represents a single tracked user (anonymous, simple followers, sporadic consumers, or loyal customers), while all the related actions are represented by the tensor A_i (note that A_i is a sub-tensor of A and is in turn a tensor).

As aforementioned, C_i represents the data structure containing the data identifying the user i (*e.g.*, personal data such as her email addresses, web pages visited, communities attended, etc.); the ability to identify the user is used to implement marketing actions towards her (for example, to send an email to a user we need to know her email address otherwise the marketing action cannot be performed).

A_i represents the tensor of all the actions registered for the user i (see Fig. 1) and is the most interesting element of the database, because it enables the possibility of doing automated data analysis and studying the evolution of the data over time. A_i can be defined as follows:

$$A_i = \{t_{i,j}, a_{i,j}, p_{i,j}\} \, j > 0 \tag{1}$$

In the formula (1) we define the actions of each user as a (potentially infinite) sequence of triplets stored as a tensor. Each triplet contains the time $t_{i,j}$ when the action happened, the type $a_{i,j}$ that characterises the action and the payload $p_{i,j}$ that is the content of the action. For each i and j we have that $a_{i,j} \in T$ where T is a finite set of possible action types, while the payload is in general a structured content where the structure depends on the action type.

The action type is needed to distinguish between the several possible actions and to characterise the content of the payload. The most typical action types are social likes, web pages opened and visited, subscription to newsletters, products/services bought from the e-commerce website, newsletters opened and read, and so on.

Focusing on what happens when a user makes an action monitored by the MA software, let us suppose that a particular user C_k adds a *like* on a post present on a social page, and suppose that $|A_k| = m$, that is, we have m triplets in A_k. When the model imports the data, the model detects the new action and registers it: a triplet $\{t_{k,m+1}, a_{k,m+1}, p_{k,m+1}\}$ is added to A_k where $t_{k,m+1}$ is the time when the user added her *like*, $a_{k,m+1}$ is the action type (in this case a "social *like*"), and $p_{k,m+1}$ is the payload. The payload content is a specific structured data type, different for each type of action and contains specific details of the action. In this example of social *likes*, the typical payload will contain, among other data, the textual content of the post which received *likes* with their social hashtags. In the long run, our model accumulates a long series of actions for each tracked user. All these data for each user i include a chronological order given by the $t_{i,j}$ and can be filtered by the action type. Moreover, as mentioned above, the payload content can be used for deep contextual analysis, because contains information about each action (*e.g.*, product ordered, web pages visited, content of the social post with hashtags) and it is useful to do semantic analysis.

As discussed in Sect. 3, the AI algorithms can access the multidimensional data using appropriate operators. The most common one is the *selection operator*, that is used to filter over the data to obtain the interesting actions for the specific research question. The *selection operator* is defined, as in relational algebra, as follows:

$$A_i' = \sigma_C(A_i) \tag{2}$$

Where σ is the selection operator, C is the logical expression regarding the filtering condition on A_i, and A_i' is the result of the selection. The logical condition can be expressed over A_i in order to select a specific set of actions, in a specified time frame with specific action types and payloads. Several kind of analysis can be carried over this large database. The following sections will show the general outlines of the natural experiment nearing completion that we are currently conducting with these type of data, while Fig. 3 shows the general architecture of our novel framework: the model proposed by us and the interaction with the existing MA software.

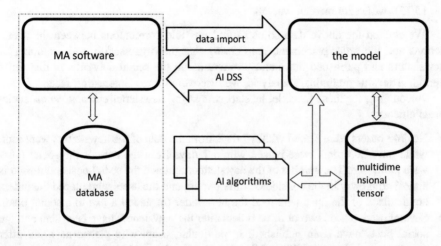

Fig. 3. The general architecture and interaction between the marketing automation software, our model, and the AI algorithms implementable into the model itself. The MA components are outside the model and are enclosed in the dashed rectangle (left side of the figure).

5   Towards a Natural Experiment

In line with Varian [4] and Rosenzweig and Wolpin [16], we started working on a first real large database originated from an e-commerce organisation that makes use of public social and editorial media content to promote its products/services with the main objective of increasing sales. The experimental subjects are the final users who buy products. The organisation sales its products/services by using an e-commerce platform. This platform is not a generalist website, rather it is specialised on a specific category of automotive products/services, so as we can assume that all the experimental subjects are similar regarding the supply they are interested in (so are quite homogeneous in this respect). All the data has been collected from the MA software and imported in our model, and users have been anonymised, so that we can distinguish between them but cannot identify them nominally. Focusing on the large database obtained, we are interested in studying the correlation between the marketing operations and the order placement.

The time-period considered in the large database is from 01 January 2017 up to 20 February 2018 (almost fourteen months). The statistical data are as follows:

- The total number of experimental subjects were 2716. Our design allows us consider as experimental subjects all the users that placed an order on the platform at least once in the aforementioned time-period.
- The number of orders placed in the considered time-period was 3861:

- Each user placed between 1 and 27 orders, and 2059 users placed only one order.
- For each user we have registered a number of actions, the mathematical average of the registered actions for each user was 36 (with a standard deviation equal to 15,75), while the median was 38.

We studied the above data so as to find possible correlations between the users' actions and the order placement, considering that the organisation sent a number of newsletters and published news and promotions on the social networks in that time-period. After our preliminary analysis, we discovered a few interesting facts.

Accounting for these facts, let us start comparing newsletter efficacy *versus* social posts efficacy:

- No. 743 orders were placed within 7 days from the date of the newsletters were sent, while only 29 orders were placed within 7 days from the date of the social posts were published. The number of the newsletters sent and the social posts published is almost equal within the considered time-period, and the users who placed the orders are a subset of the ones who read the newsletter (or added a *like* to a social post).
- Several users placed two or more orders after the newsletters have been sent and the social posts have been published. In particular, users who placed an order after opening a newsletter are 590, while only 27 placed an order after adding a *like* on a social post.

In other words, the orders placed increase significantly after sending newsletters to the users, while the order increase is very modest as a consequence of social network activities. We do not intend to argue that, generally, it is better to send newsletters than publishing posts on social networks but, in this particular case, the database supports this evidence. Therefore, we started studying more deeply the action made by the users (limited to the users who reacted positively to the newsletter) and the conversion ratio c_N of each newsletter, that is, the ratio of the number of users who placed an order after receiving a newsletter *versus* the total number of users who received the newsletter:

$$c_N = \frac{o_N}{s_N} \quad \quad (3)$$

In the formula (3), N is a generic newsletter, c_N is the conversion ratio of the newsletter N, o_N is the number of orders placed and correlated to the newsletter N, and s_N is the number of newsletters sent (thus the number of emails sent) to the users. This requires a few explanations. Particularly, we are referring to the total number of emails sent, which is different from the number of emails read/opened by the users. To cope with this task, let us call r_N the number of emails effectively read by the users. In this respect, we can calculate the efficacy e_N of a newsletter as the ratio of the number of emails read *versus* the total number of emails sent:

$$e_N = \frac{r_N}{s_N} \quad \quad (4)$$

It is intuitive as well as supported by the data that the number of read emails is related to the number of orders placed. Therefore, in order to increase c_N, the experimenters have two possibilities: the first is to increase the number of orders, while the second, albeit apparently paradoxical, is to reduce s_N.

The first experimenter's strategy is related to the increase of r_N: if the experimenters were able to increase the number of users that read the newsletters, the experimenters should have an effect on o_N (more final users reading the newsletter should reflect on more orders placed). The second experimenter's strategy is only apparently paradoxical, because if we were able to reduce s_N without reducing r_N, we will obtain a more effective newsletter, addressing only users willing to open it and reducing the risk of being considered like spam from users not interested in the newsletters.

Arguably, the first strategy has an emergent tangible economic value: in fact, more orders placed by the users means more revenues for the organisation. The second strategy, conversely, has an intangible value consisting in reducing the risk of being considered unsolicited or undesired (*e.g.*, electronic spamming), and thus no emergent tangible value. With regard to the first strategy, we designed the following experimental treatment: the users are divided in two groups, U_1 and U_2, and they are random assigned to U_1 and U_2, where U_1 is the treatment group. For each user $u \in U_1$ the MA software sends newsletters at a specific time t_u, calculated as the time when the user is presumed to be active on internet (based on reading time of earlier newsletters or activity time on social networks and e-commerce website recovered from the action list). We are working in order to find one or more algorithms to assign the sending time t_u for each user. The second group U_2 is our control group: the newsletters are sent with the usual strategy (the date and time of day decided by the e-commerce firm). We will compare the conversion ratio of the first group with the conversion ratio of the baseline group so as to measure the effectiveness of this strategy. With regard to the second strategy, we have several ideas to find the users that should not receive the newsletter (for example, using Recurrent Neural Networks, as outlined in the next section). The natural experiment is designed to compare the result of each newsletter sent and opened with the forecast made by an algorithm on which we are working on: basically, the efficacy of our algorithm is given by comparing the set of users that opened each newsletter with the set of users forecasted by our algorithm. For each newsletter N, ideally, the algorithm should be able to reduce to zero both the wrong positive forecasts W_N^+ (defined as the set of users not included in the list that in the real case opened the newsletter) and the wrong negative forecasts W_N^- (the set of users forecasted as newsletter openers that in the real case did not open the newsletter). In formula, let O_N be the set of users that opened the newsletters and F_N the set forecasted by the second strategy. Using a mathematical set notation, the wrong forecasts can be calculated as a set difference, where $\setminus$ is the notation for set subtraction:

$$W_N^+ = O_N \setminus F_N \tag{5}$$

$$W_N^- = F_N \setminus O_N \tag{6}$$

Our objective is to minimise the size of both forecasts:

$$\min\left(\left|W_N^+\right| + \left|W_N^-\right|\right) \tag{7}$$

By using the first strategy, we cannot know in advance if and how much the strategy will be effective: only when we will complete the natural experiment with future newsletters we will be able to know the efficacy and eventually tune it to be more effective. Pursuing the second strategy, we are able to test the second strategy on existing data to tune the algorithm – and then experiment – on future newsletters, having the calculated efficacy of historical data as a baseline.

6 Next Steps to Be Implemented

As we argued in Sect. 5, we started to conduct a natural experiment, which is nearing completion, by using real input data of final users with the MA software. Our first step is about investigating the different conversion ratio of newsletters (where the conversion ratio is the ratio between the number of users that placed an order *versus* the number of users that received the newsletter) in order to find a way to increase the conversion ratio. In the near future, our research agenda will focus (i) on developing a microeconomic theoretical framework related to the consumption setting investigated in this paper, and (ii) on completing the natural experiment started with two different strategies: the first is to send the newsletter at different times during the day, because our data analysis over the activities extracted from the multidimensional data shows that different users read the e-mail at different times. In this respect, we are collecting further data to verify the effectiveness of this strategy, that is based on a simple statistical correlation. Following this strategy, we do not reduce the number of email sent but try to adapt to the preferred email opening time of users. The second strategy is about reducing the number of emails sent by selecting the most likely users that will open the newsletter. We are experimenting it also with Recurrent Neural Networks (RNN: for an overview see [13]) that seems the most effective to model along the time dimension and arbitrary sequence of events and inputs. Even in this case, we are collecting big data to have a statistical significance about the results, that seems quite promising. Our main objective with the model presented in this paper is to provide an automatic tool that meets the typical research questions within organisations using marketing automation software. As long as we find good strategies, the model will be able both to help decisions and provide automatic intelligent strategies that improve the effectiveness of the marketing actions. We will continue to study different algorithms and strategies as well as the natural experimentation. On a long term range, finally, our objective will be to promote our model as a decision support system for marketing automation backed by artificial intelligent algorithms equipped with a friendly and intuitive graphical user interface. To reach this long term goal, we will continue to study several algorithms and we plan to add semantic analysis of the content so as to improve the effectiveness of deep learning algorithms and broaden the possibilities of the MA software.

References

1. Chen, S.-H., Kao, Y.-F., Venkatachalam, R.: Computational behavioral economics. In: Frantz, R., Chen, S.-H., Dopfer, K., Heukelom, F., Mousavi, S. (eds.) Routledge Handbook of Behavioral Economics, pp. 297–315. Routledge, Abingdon, Oxon (2017)
2. Kao, Y.-F., Velupillai, V.K.: Behavioural economics: classical and modern. Eur. J. Hist. Econ. Thought 22(2), 236–271 (2015)
3. Dell'Aversana, R.: A unified framework for multicriteria evaluation of intangible capital assets inside organizations. In: Bucciarelli, E., Chen, S.-H., Corchado, J.M., (eds.) Decision Economics: In the Tradition of Herbert A. Simon's Heritage, pp. 114–121. Springer, Cham (2018)
4. Varian, H.R.: Causal inference in economics and marketing. Proc. Nat. Acad. Sci. U.S.A. (PNAS) 113(27), 7310–7315 (2016)
5. Einav, L., Levin, J.: Economics in the age of big data. Science 346(6210), 1243089 (2014)
6. Simon, H.A.: Administrative Behavior: A Study of Decision-Making Processes in Administrative Organizations, 4 th edn. Free Press, New York (1957, 1976, 1997) [1945]
7. Varian, H.R.: Big data: new tricks for econometrics. J. Econ. Perspect. 28(2), 3–27 (2014)
8. Chen, S.-H. (ed.): Genetic Algorithms and Genetic Programming in Computational Finance. Kluwer, New York (2002)
9. Goldberg, D.E.: Genetic Algorithms in Search, Optimization, and Machine Learning. Addison-Wesley, Reading (1989)
10. Mitchell, M.: An Introduction to Genetic Algorithms. M.I.T. Press, Cambridge (1996)
11. Chen, S.-H., Kuo, T.-W., Shien, Y.-P.: Genetic programming: a tutorial with the software simple GP. In: Chen, S.-H. (ed.) Genetic Algorithms and Genetic Programming in Computational Finance, pp. 55–77. Kluwer, New York (2002)
12. Deng, L.: A tutorial survey of architectures, algorithms, and applications for deep learning. APSIPA Trans. Signal Inf. Process. 3, E2 (2014)
13. Schmidhuber, J.: Deep learning in neural networks: an overview. Neural Netw. 61, 85–117 (2015)
14. Marlow, S. (ed.): Haskell 2010 language report (2010). https://www.haskell.org/onlinereport/haskell2010. Accessed 2 Feb 2018
15. Odersky, M., Spoon, L., Venners, B.: Programming in Scala: Updated for Scala 2.12, 3rd edn. Artima Press, Walnut (2016)
16. Rosenzweig, M.R., Wolpin, K.I.: Natural "natural experiments" in economics. J. Econ. Lit. 38(4), 827–874 (2000)

Google Trends and Cognitive Finance: Lessons Gained from the Taiwan Stock Market

Pei-Hsuan Shen, Shu-Heng Chen[✉], and Tina Yu

National Chengchi University AI-ECON Research Center,
Taipei, Taiwan, Republic of China
chen.shuheng@gmail.com

Abstract. We investigate the relationship between Google Trends Search Volume Index (SVI) and the average returns of Taiwan Stock Exchange Capitalization Weighted Stock Index (TAIEX). In particular, we used the aggregate SVI searched by a company's abbreviated name and by its ticker symbol to conduct our research. The results are very different. While the aggregate SVI of abbreviated names is significantly and positively correlated to the average returns of TAIEX, the aggregate SVI of ticker symbols is not. This gives strong evidence that investors in the Taiwan stock market normally use abbreviated names, not ticker symbols, to conduct Google search for stock information. Additionally, we found the aggregate SVI of small–cap companies has a higher degree of impact on the TAIEX average returns than that of the mid–cap and large–cap companies. Finally, we found the aggregate SVI with an increasing trend also has a stronger positive influence on the TAIEX average returns than that of the overall aggregate SVI, while the aggregate SVI with a decreasing trend has no influence on the TAIEX average returns. This supports the attention hypothesis of Odean [12] in that the increased investors attention, which is measured by the Google SVI, is a sign of their buying intention, hence caused the stock prices to increase while decreased investors attention is not connected to their selling intention or the decrease of stock prices.

Keywords: Google Trends · Investors attention · TAIEX
Cognitive finance · Search volume index · Attention hypothesis
Stock returns

1   Introduction

Investors' attention plays an important role in their buying decisions [12] and in stock pricing [11]. When buying a stock, investors are faced with a large number of choices. Since human beings have bounded rationality [14], the cognitive and temporal abilities of an investor to process stocks related information are limited. To manage this problem of choosing among many possible stocks to purchase,

© Springer Nature Switzerland AG 2019
E. Bucciarelli et al. (Eds.): DCAI 2018, AISC 805, pp. 114–124, 2019.
https://doi.org/10.1007/978-3-319-99698-1_13

Odean [12] proposed that investors limit their choices to stocks that have recently caught their attention. When selling stocks, however, attention has no effect, because investors tend to sell stocks that they own. In [3], Barber and Odean validated this attention hypothesis empirically using indirect attention measures, including news, high abnormal trading volume, and extreme one–day returns.

When attention–grabbing stocks are the subjects of buying interests, the buying pressure would drive these stocks' prices upward. Using Google Trends Search Volume Index (SVI) as a direct measure of investors' attention, Da, Engelberg and Gao [7] sampled Russell 3000 stocks and found that the increase in the SVI could predict higher stock prices in the short term and price reversals in the long run. The positive correlation between Google SVI and stock returns has also been observed in the S&P 500 stocks [8], and the stocks traded in the German [2] and the Japan [16] stock markets.

In this paper, we investigate the relationship between Google SVI and the average returns of *Taiwan Stock Exchange Capitalization Weighted Stock Index (TAIEX)*. Unlike previous works which first modeled the relationship between Google SVI and an individual stock's returns and then aggregated the results, we first aggregated the SVI of all companies included in TAIEX and then modeled the relationship between the aggregate SVI and the TAIEX average returns. Since TAIEX is a capitalization weighted index, we used capitalization as the weight to aggregate the SVI of each company.

In our study, we used the aggregate SVI of a company's abbreviated name and of its ticker symbol to conduct our research. We found they give different results. The aggregate SVI of abbreviated names is significantly and positively correlated to the average returns of TAIEX, which is similar to that reported in previous works [2,7,8] and [16]. By contrast, the aggregate SVI of ticker symbols has no impact on the TAIEX average returns. This gives strong evidence that investors in the Taiwan stock market normally use abbreviated names, not ticker symbols, to conduct Google search for stock information.

Additionally, we found the aggregate SVI of small–cap companies has a higher degree of impact on the TAIEX average returns than that of the mid–cap and large–cap companies. This result is similar to that of Russell 300 stocks [7] but is different from the result of stocks traded in the German stock market [2].

Finally, we found the aggregate SVI with an increasing trend, i.e. $SVI_t > SVI_{t-1}$, has more positive impact on the TAIEX average returns than the overall aggregate SVI has. Moreover, the aggregate SVI with a decreasing trend, i.e. $SVI_t \leq SVI_{t-1}$, has no effect on the TAIEX average returns. These combined results support the attention hypothesis of Odean [12], which states that investors' attention only influences their stock buying decisions, not stock selling decisions. Using Google SVI as a proxy of investors' attention, an increased SVI is a sign of buying intention, which would lead to possible stock prices increase. By contrast, selling intention has no influence on the Google SVI, hence, a decreased SVI is not directly connected to the decrease of stock returns.

The rest of the paper is organized as follows. Section 2 summarizes related works. Section 3 explains the data and methods used to conduct our research.

The results are then presented and analyzed in Sect. 4. Finally, Sect. 5 gives our concluding remarks.

2   Related Works

Google Trends (`trends.google.com/trends/`) provides data on search term frequency dating back to January of 2004. The search frequency is normalized to an index called *Search Volume Index* (SVI) such that the highest search frequency within the search period has index value 100 and the rest of the frequencies have index between 0 and 100. To search information about a stock on Google, a user can enter either its ticker symbol or the company name. In [7], Da, Engelberg, and Gao used ticker symbols of Russell 3000 stocks as search keywords to obtain their SVIs for the period of 2004–2008. They then ran Vector Autoregression (VAR) model for each stock's *abnormal SVI* on the stock's following week *abnormal returns*. After that, they averaged the VAR coefficients across all stocks. The p–value is also computed using a block bootstrap procedure under the null hypothesis that all VAR coefficients are zero. Their results showed that the abnormal SVI can positively and significantly predict the abnormal returns over the next two weeks and the predictive power of abnormal SVI is stronger among smaller stocks.

Joseph, Babajide Wintoki and Zhang also used the ticker symbols of S&P 500 stocks to obtain their SVIs from 2005–2008 to conduct research [8]. They first sorted these SVIs into 5 portfolios, from the highest SVI to the lowest SVI. They then compared the stocks' following week's *abnormal returns* in the 5 portfolios. They found that there is a monotonic increase of abnormal stock returns from the lowest SVI portfolio to the highest SVI portfolio.

While SVI based on the ticker symbol of a stock might reveal investors' attention on the stock more closely, Bank, Larch, and Peter [2] are interested in the question of how public interest in a firm influences stock market activity. For that purpose, they studied the stocks traded in the German stock market by using their company names given by the Thomson Reuters Datastream as the search keywords to obtain their SVIs from 2004 to 2010. Using a similar portfolio–based analysis in [8] with 3 portfolios, they found a moderate relation between the *change of SVI* and the stock's next month *excess returns*. However, after incorporating market capitalization of the stock to refine the original 3 portfolios into 9 portfolios, they found that the portfolio of stocks with a large *change of SVI* and large market capitalization has much higher next month returns than that of the portfolio of stocks with a small *change of SVI* and small market capitalization.

For the Japan stock market, Takeda and Wakao [16] also used company names as keywords to obtain the SVIs of 189 stocks included in the Nikkei 225 from 2008 to 2011. They divided the SVIs into 4 portfolios using three criteria: SVI in [8], *change of SVI* in [2] and *abnormal SVI* in [7]. They observed that the *change of SVI* values can be positive or negative while the *abnormal SVI* values are more stable and smooth. Their analysis showed that under the grouping strategies of [2] and [7], the portfolio with the highest SVI has the largest next week abnormal returns.

Similar researches have also been conducted for stocks traded in other stock markets, such as NASDAQ & NYSE [18], France [1] and Turkey [17].

There are also works investigating the relationship between asset indexes performance and Google SVI, searched by the index names. For example, Voz-lyublennaia [19] studied a set of six asset indexes, including Dow Jones Industrial Average (DJIA) index, NASDAQ index, S&P 500 index, the 10 year Treasury index, the Chicago Board Options Exchange Gold index and the West Texas Intermediate crude oil index. He found the increased attention to an index has a significant short-term effect on the index's return. However, the price pressure can be either positive or negative, depending on the nature of the information uncovered by the Google search. This result is different from that of previously mentioned works [2,7,8,16], which endorsed Odean's attention hypothesis that retail investors are more likely to buy than sell a security that attracts their attention, hence investors' attention normally creates positive price pressures.

Another work is by Latoeiro, Ramos and Veiga [10], who studied the EURO STOXX 50 index performance related to the Google SVI searched by EURO STOXX. Their results show that an increase in SVI for the index predicts a drop in the market index, which is different from that of the U.S. market indexes reported in [19]. Also, the SVI with an increasing trend is statistically significant in impacting the index returns but the SVI with a decreasing trend is not.

In addition to stock prices, Google SVI has also been used to predict the prices of digital currencies. In [9], Krištoufek reported that there is a very strong bidirectional positive correlation between the price of BitCoin and the SVI searched by "BitCoin". He found that when the interest in the BitCoin currency, measured by the Google SVI, increases, so does its price. Similarly, when the BitCoin price increases, it generates more interest of the currency not only from investors but also from the general public. This is not surprising since there is no macroeconomic fundamentals for the digital currency and the market is filled with short-term investors, trend chasers, noise traders and speculators. Additionally, it is quite easy to invest in BitCoin as the currency does not need to be traded in large bundles. Consequently, the Google SVI of the digital currency influences the price of the digital currency and vice visa. However, these bidirectional effects are short-lived for two periods (weeks) only.

Google SVI has also been used to build currency exchange rate models to perform forecasting. The key ingredients of these models are macroeconomic fundamentals, such as inflation, which are normally released by government with a monthly time lag. In [4], Bulut used Google SVI of related keywords to now-cast these fundamentals to built two currency exchange rate models. The results indicate that inclusion of the Google Trends-based nowcasting values of macro fundamentals to the current set of government released-macro-economic variables improve the out-of-sample forecast of Purchasing Power Parity model in seven currency pairs and of Monetary model in four currency pairs.

In [13], Preis, Moat & Stanley incorporated Google search volume to devise the following trading strategies:

```
if ΔSVI(t − 1, Δt) > 0
      sell at the closing price p(t) on the first
      trading day of week t and buy at price p(t − 1) at the end
      of the first trading day of the following week
if ΔSVI(t − 1, Δt) < 0
      buy at the closing price p(t) on the first trading day
      of week t and sell at price p(t+1) at the end of the
      first trading day of the coming week
```

where $\Delta SVI(t-1, \Delta t) = SVI(t-1) - MA_{SVI}(t-2, \Delta t)$, $MA_{SVI}(t-2, \Delta t)$ is the Δt weeks moving average of Google SVI between weeks $t-2$ and $t-2-\Delta t$.

They tested the strategies using a set of 98 search keywords on the DJIA index from 2004 to 2011 under $\Delta t = 3$. They found that the overall returns from the strategies are significantly higher than the returns from the random strategies. Among them, the SVI of the search keyword *debt* gives the best performance of 326% profit, which is much higher than the 33% profit yield by the historical pricing strategy (replacing SVI with the DJIA prices in the above strategies) and the 16% profit produced by the "buy and hold" strategy.

The predictive power of Google Trends data for the future stock returns has also been challenged. In [6], Challet and Ayed applied non-linear machine learning methods and a backtest procedure to examine if the Google SVI data contain more predictive information than the historical price returns data. They downloaded SVI data searched by company tickers and names from 2004 to 2013-04-21. They also obtained historical pricing data for the same testing period. After processing the two sets of data, their backtest system shows that both data give similar accumulative returns, after transaction costs. The authors believe that SVI data share many similar properties with the price returns: (1) both are aggregate signals created by many individuals; (2) they reflect something related to the underlying assets, (3) both are very noisy. Consequently, the backtest system found them contain about the same amount of predictive information.

3   Research Methods

3.1   TAIEX Weekly Average Open Prices

TAIEX is the capitalization–weighted index of companies that are traded in the *Taiwan Stock Exchange (TWSE)*. From the website of *Taiwan Economic Journal (TEJ)*, a database that contains historical financial data and information of the major financial markets in Asia, we downloaded TAIEX *weekly average opening prices* between January 5, 2014 and November 6, 2016.

3.2   Aggregate Google SVI Variable

Two sets of SVI data were downloaded from Google Trends using two sets of search terms. The first set consists of the *abbreviated names* of 849 companies traded on TWSE and the second set contains the *ticker symbols* of these companies. The following subsections explain the data processing procedures.

Abbreviated Names. A stock traded on TWSE has an abbreviated name to represent the company. For example, 台積電 is the abbreviated name for 台灣積體電路製造股份有限公司. We retrieved the SVI in the Taiwan region using the abbreviated name of each company traded on TWSE from January 5, 2014 to November 6, 2016. However, we found some small-cap companies have some weekly SVI data missing. In addition, some abbreviated names are common terms that may be used by non-investors to conduct Google search for non-investment related information. In these two situations, we replaced the search results with the results obtained using their ticker symbols. The total number of stocks whose SVI have been replaced under this process is 49.

To aggregate the 849 SVIs into a single index, we used a weighted sum approach, where the weight is the company size, represented by its relative percentage of market value on November 18, 2016. The information was obtained from the website of *Taiwan Futures Exchange*. This approach is based on the following assumptions:

- Each search volume is independent. Increased attention on one stock will not influence others.
- The higher a company's market value is, the more attention the company receives and hence the higher the search volume.
- The companies that constitute TAIEX remain unchanged.

The aggregate SVI time series contains 146 weeks of data.

Ticker Symbols. A stock traded on TWSE also has a ticker symbol. For example, the ticker symbol of 台積電 is 2330. We first used the ticker symbol of each stocks to retrieve their SVIs. Next, we used the same procedures described in the previous section to obtain the aggregate SVI. The time series also has 146 weeks of data.

3.3   Econometric Method

Newey–West correction of standard error is a method to estimate the coefficients of a linear regression model applied to time series data. It is used to correct *autocorrelation* (also called serial correlation) and *heteroskedasticity* in the error terms in the regression model. We applied the method implemented in the statistical software SAS to generate our linear regression models.

Following [7], we first converted all time series data into natural logarithm (ln). In this way, coefficients on the natural–log scale are directly interpretable as approximate proportional changes. For example, with a coefficient of 0.06, a change of 1 in the independent variable corresponds to an approximate 6% change in the dependent variable. Moreover, the transformation reduces the scale difference of the variables, hence increases model prediction accuracy.

The linear regression model is as follows:

$$R_t = \beta_0 + \beta_1 \Delta svi_{t-1} + \epsilon_t \tag{1}$$

where $R_t = ln(p_t) - ln(p_{t-1})$, p_i is the TAIEX price on week i, and $\Delta svi_{t-1} = ln(SVI_{t-1}) - ln(SVI_{t-2})$, which is the *change of SVI* in [8]. We used the aggregate SVI of abbreviated names and of ticker symbols to run the regression. The results are compared in Sect. 4.1.

Next, we are interested in knowing if the Δsvi of companies with different capitalization has a different degree of impact on the TAIEX average returns. To answer that question, we ran the following four linear regression models:

$$\begin{aligned}
R_t &= \beta_0 + \beta_1 \Delta svi_{large,t-1} + \epsilon_t \\
R_t &= \beta_0 + \beta_2 \Delta svi_{middle,t-1} + \epsilon_t \\
R_t &= \beta_0 + \beta_3 \Delta svi_{small,t-1} + \epsilon_t \\
R_t &= \beta_0 + \beta_4 \Delta svi_{rest,t-1} + \epsilon_t
\end{aligned} \tag{2}$$

where the subscript *large* stands for *large–cap* (top-50 companies), *middle* stands for *mid–cap* (top–51 to top–150 companies), *small* stands for *small–cap* (top–151 to top–450 companies) [15] and *rest* stands for the rest 399 companies. We also used both aggregate SVI of abbreviated names and of ticker symbols to run the regression. The results are presented in Sect. 4.2.

According to Odean [12], investors' attention only impacts their stock buying decisions, not stock selling decisions. Using Google SVI as a proxy of investors' attention, this means that an increased SVI is a sign of buying intention, which leads to possible stock prices increase. By contrast, selling intention has no impact on the Google SVI. Hence, a decreased SVI is not directly connected to the decrease of stock returns.

To test this hypothesis, we divided the aggregate SVI into two groups: one with an increasing trend, i.e. $SVI_t > SVI_{t-1}$, and the other with a decreasing trend, i.e. $SVI_t \le SVI_{t-1}$. We then used the two aggregate SVIs to run the linear regression model of Eq. 1. The results are analyzed in Sect. 4.3.

4   Results and Analysis

4.1   Abbreviated Names vs. Ticker Symbols

Table 1 shows that the Δsvi of abbreviated names is statistically significant in impacting TAIEX average returns while the Δsvi of ticker symbols is not. This suggests that investors in the Taiwan stock market normally use abbreviated names, not ticker symbols, to conduct Google search for stock information. This makes sense, as the ticker symbols of Taiwanese stocks are 4–digit numerical values, which could be confused as product numbers, specific year, phone extension or other meaning by the Google search engine, hence produces irrelevant search results. By contrast, abbreviated names are less ambitious and are easily linked to the company stocks that a Google user is searching for. Consequently, investors are more likely to use abbreviated names, rather than ticker symbols, to search for stock information to obtain relevant results. This discovery of investors' behaviors supports the belief that Google search data have the potentials to reveal people's interests, intentions and possible future actions [5].

Table 1. Δsvi (Abbreviated Names & Ticker Symbols) on TAIEX Average Returns

	Abbreviated Names				Ticker Symbols			
Parameter	Estimate	Std Err	t	p-value	Estimate	Std Err	t	p-value
Intercept	0.001057	0.000646	1.64	0.1039	0.001099	0.000652	1.69	0.0940
Δsvi	0.037486	0.0186	2.02	0.0454*	0.011643	0.0136	0.86	0.3922

Notes: $*\,*\,*$, $**$ and $*$ indicate statistical significance at the 0.1%, 1% and 5% levels.

The Δsvi of abbreviated names is positively related to the TAIEX average returns, which is similar to that reported in previous works [2,7,8] and [16].

4.2 Large vs. Middle vs. Small Capitalization

Table 2 shows the aggregate SVI of *abbreviated names* for *large–cap*, *mid–cap* and *small–cap* companies are all significant in impacting the TAIEX average returns. Additionally, their coefficients show the aggregate SVI of *small–cap* companies has a larger impact on the TAIEX average returns than that of the *mid–cap* and the *large–cap* companies: increasing the value of Δsvi_{small}, Δsvi_{mid} and Δsvi_{large} by 1 will increase the TAIEX average returns by 0.075961, 0.072652 and 0.062965 respectively. This result is similar to that of the Russell 300 stocks [7] but is different from the stocks traded in the German stock market [2].

Table 2. Δsvi for Companies of Different Capitalization on TAIEX Average Returns

	Abbreviated Names				Ticker Symbols			
Parameter	Estimate	Std Err	t	p-value	Estimate	Std Err	t	p-value
Intercept	−0.00087	0.00102	−0.85	0.3982	0.000398	0.00106	0.38	0.7074
Δsvi_{large}	0.062965	0.0125	5.05	<.0001***	0.028567	0.0104	2.74	0.0079**
Intercept	−0.00051	0.00123	−0.41	0.6813	0.000661	0.00134	0.49	0.6231
Δsvi_{mid}	0.072652	0.0202	3.60	0.0006***	0.014188	0.0208	0.68	0.4978
Intercept	−0.00126	0.00120	−1.05	0.2969	−0.00063	0.00125	−0.50	0.6157
Δsvi_{small}	0.075961	0.0226	3.36	0.0013**	0.027574	0.0139	1.99	0.0502
Intercept	0.001918	0.00168	1.14	0.2578	0.002607	0.00147	1.78	0.0802
Δsvi_{rest}	0.048907	0.0440	1.11	0.2705	0.008281	0.0243	0.34	0.7345

Notes: $*\,*\,*$, $**$ and $*$ indicate statistical significance at the 0.1%, 1% and 5% levels.

The aggregate SVI of *ticker symbols* for the *large–cap* companies is also significant in impacting the TAIEX average returns, although the degree (coefficient)

of its impact is much lower than that of the aggregate SVI of *abbreviated names*. This means that investors also use the ticker symbols of large–cap companies to conduct Google search for large–cap companies stock information. This also makes sense because stocks of large–cap companies are traded more often; hence their ticker symbols are easily associated with their companies by the Google search engine to generate relevant search results. Furthermore, investors tend to remember the tickle symbols of more frequently traded stocks. These explain why the aggregate SVI of ticker symbols for the large–cap companies is significantly correlated to the TAIEX average returns.

4.3 Validation of the Attention Hypothesis

Section 4.1 shows the aggregate SVI of abbreviated names is positively and significantly correlated to the TAIEX average returns. In this section, we used this SVI data to validate the attention hypothesis of Odean [12]. As shown in Table 3, there are 66 weeks of data in this aggregate SVI that have an increasing trend. Similar to the entire 146 weeks of data, these increasing trend data are also positively and significantly correlated to the TAIEX average returns. However, the increasing trend data have more positive impact (larger coefficient) and more significant impact (smaller p–value) on the TAIEX average returns. By contrast, the 80 weeks of decreasing trend data have no impact on the average returns of TAIEX. The combination of these results supports the attention hypothesis of Odean [12] in that increased investors' attention, which is measured by the Google SVI, is connected to the increased stock prices while decreased attention is not connected to the decrease of stock prices.

Table 3. Δsvi with Increasing/Decreasing Trends on TAIEX Average Returns

Parameter	Increasing Trend Data (66 weeks)				Decreasing Trend Data (80 weeks)			
	Estimate	Std Err	t	p-value	Estimate	Std Err	t	p-value
Intercept	−0.00237	0.00282	−0.84	0.4045	0.004747	0.00269	1.76	0.0818
Δsvi	0.083459	0.0253	3.29	0.0016**	0.037635	0.0244	1.54	0.1265

Notes: ∗ ∗ ∗, ∗∗ and ∗ indicate statistical significance at the 0.1%, 1% and 5% levels.

5 Concluding Remarks

Google Trends data have been linked to various economic indicators, including automobile sales, unemployment claims, travel destination planning and consumer confidence [5]. In this study, we investigate the relationship between Google Trends SVI and the average returns of TAIEX. In addition to identifying their significant and positive correlation, similar to that reported in previous works, we also discover that Taiwan investors normally use a company's abbreviated name, rather than its ticker symbol, to conduct Google search for stock related information. We will continue exploring other investors' buying and selling intentions/behaviors by evaluating expanded Google search keywords using other tools such as Google Correlate.

Google SVI of small–cap companies is found to have a stronger impact on the TAIEX average returns than that of the mid–cap and the large–cap companies. This result is similar to that of the Russell 300 stocks but is different from the stocks traded in the German stock market. Why this difference? Is it due to the differences of the two different stock market structures or is it due to other factors? We will address this question in the future.

Acknowledgement. The authors are grateful for the research support in the form of Ministry of Science and Technology (MOST) Grants, MOST 106-2410-H-004-006-MY2.

References

1. Aouadi, A., Arouri, M., Teulon, F.: Investor attention and stock market activity: evidence from France. Econ. Modell. **35**, 674–681 (2013)
2. Bank, M., Larch, M., Peter, G.: Google search volume and its influence on liquidity and returns of German stocks. Financ. Mark. Portf. Manag. **25**(3), 239–264 (2011)
3. Barber, B.M., Odean, T.: All that glitters: the effect of attention and news on the buying behavior of individual and institutional investors. Rev. Financ. Stud. **21**(2), 785–818 (2008)
4. Bulut, L.: Google Trends and forecasting performance of exchange rate models. IPEK Working Papers 1505, Ipek University, Department of Economics (2015)
5. Choi, H., Varian, H.: Predicting the present with Google Trends. Econ. Rec. **88**, 2–9 (2012)
6. Challet, D., Ayed, A.B.H.: Do Google Trend data contain more predictability than price returns? J. Invest. Strat. (2015)
7. Da, Z., Engelberg, J., Gao, P.: In search of attention. J. Financ. **66**(5), 1461–1499 (2011)
8. Joseph, K., Wintoki, M.B., Zhang, Z.: Forecasting abnormal stock returns and trading volume using investor sentiment: evidence from online search. Int. J. Forecast. **27**(4), 1116–1127 (2011)
9. Krištoufek, L.: BitCoin meets Google Trends and Wikipedia: Quantifying the relationship between phenomena of the Internet era. Scientific Reports 3 (2013). Article number 3415
10. Latoeiro, P., Ramos, S.B., Veiga, H.: Predictability of stock market activity using Google search queries. Working Paper 13-06. Universidad Carlos III de Madrid (2013)
11. Merton, R.C.: A simple model of capital market equilibrium with incomplete information. J. Financ. **42**(3), 483–510 (1987)
12. Odean, T.: Do investors trade too much? Am. Econ. Rev. **89**(5), 1279–1298 (1999)
13. Preis, T., Moat, H. S., Stanley, H. E.: Quantifying trading behavior in financial markets using Google Trends, Science Report 3 (2013)
14. Simon, H.A.: A behavioral model of rational choice. Q. J. Econ. **69**(1), 99–118 (1955)
15. Taiwan Stock Exchange: TWSE Publishes the SmallCap 300 Sub-Index (2015). (Press Release). http://www.twse.com.tw/en/news/newsDetail/72DAE 07E05F945A8953D40A303F418B7

16. Takeda, F., Wakao, T.: Google search intensity and its relationship with returns and trading volume of Japanese stocks. Pac.-Basin Financ. J. **27**, 1–18 (2014)
17. Turan, S.S.: Internet search volume and stock return volatility: the case of Turkish companies. Inf. Manag. Bus. Rev. **6**(6), 317–328 (2014)
18. Vlastakis, N., Markellos, R.N.: Information demand and stock market volatility. J. Bank. Financ. **36**(6), 1808–1821 (2012)
19. Vozlyublennaia, N.: Investor attention, index performance, and return predictability. J. Bank. Financ. **41**, 17–35 (2014)

Research on the Evaluation of Scientists Based on Weighted h-index

Guo-He Feng[1](✉) and Xing-Qing Mo[2]

[1] Scientific Laboratory of Economic Behaviors, South China Normal University,
Guangzhou 510006, Guangdong, China
ghfeng@163.com
[2] School of Economics and Management, South China Normal University,
Guangzhou 510006, Guangdong, China

Abstract. [Purpose/meaning] Solving the problem that h-index and h-type index lack of comprehensive evaluation of scientists' overall academic contribution. [Method/process] Considering the contribution of scientists' first h academic papers, this paper proposed two weighted h-index model which are named h_w-index and h_{w_t}-index, and then used the data from 30 active Chinese scholars in Library and Information Science field for empirical analysis. [Results/conclusion] Revealing highly cited papers and considering the contribution of scientists' every paper, h_w-index not only weakens the influence of self-citation on the results, but also makes it easy to distinguish scientists' contributions. The h_{w_t}-index focuses on the scientists' papers in recent years, and it also considers their past contributions. Therefore, in short term evaluation, the h_{w_t}-index is more reasonable for young scientists who have made a great contribution in the past years. Potential scholars can be identified by the way of comparing h_w-index and h_{w_t}-index.

Keywords: h-index · Cited frequency · h_w-index · h_{w_t}-index

1 Research Background

In the past, the number of published papers and the citation frequency were used to evaluate the academic level and influence of scholars. The two indexes are reasonable and their calculation method is very simple, however, they also have shortcomings, such as they can't obviously reflect the academic level of scholars (Sui 2013). Hirsch (2005) proposed the h-index, which is used to evaluate the academic contribution of scholars, and it takes into account the quality and quantity of the academic output. h-index means that a scholar who has published N papers has h papers which are at least cited h times, while the remaining N - h papers are each cited less than or equal to h.

However, the h-index also has the following disadvantages: (1) The value of h-index keeps rising, which is unfair for young scholars (Rodrigo 2007). (2) The h-index values of some scholars are common and it cannot be distinguished from researchers with the same h-index (Wang et al. 2011). (3) The level of scholar's partner also has an influence on the h index, and h-index does not reflect the role of individuals in teamwork (Yu and Wang 2017; Wang et al. 2011). (4) h-index has no comprehensive

E. Bucciarelli et al. (Eds.): DCAI 2018, AISC 805, pp. 125–133, 2019.
https://doi.org/10.1007/978-3-319-99698-1_14

studies on the actual contribution of each paper, and it is not sensitive to highly cited papers, which is unfair for scholars who are highly cited but only have few papers. This paper attempts to improve the h-index to overcome the above problems.

2 Literature Review

The h-index is simple to calculate and can reflect the academic level of scholars. In recent years, it has been widely studied by researchers, and a number of h-type indexes have emerged.

However, h-index is not sensitive to highly cited papers, Egghe (2006) proposed the g-index, which means that N papers published by a scholar are ranked by the frequency of citation from high to low, and the accumulated citation frequency of the previous g papers is no less than g^2, while the cumulative frequency of the former $g + 1$ papers is less than $(g + 1)^2$. The g-index highlights the contribution of highly cited papers in the evaluation of scholars, which is fairer to scholars who have published fewer papers but are cited more frequently (Sui 2013). In addition, Zhang (2009) proposed the e-index, which is the square root of the consequence that the cumulative citation frequency of the previous h papers minus h-index squared, while the N papers published by a scholar are ranked by the frequency of citation from high to low. Based on the h-index, the w-index proposed by Wu (2010) also considers the importance of highly cited paper, and w-index means that a scholar with N papers published has w papers which are at least cited 10 w times, while the remaining w + 1 articles are cited less than 10 (w + 1) times. However, the g-index, e-index and w-index do not consider the low cited papers, which, as a result, do not fully reflect the scientific research work and academic level of scholars.

For the problem that the h-index of a scholar is affected by the level of his partner, Shekofteh (2009) proposed the Y-index, considering the contribution of the first author and the corresponding author. Based on the h-index, Hirsch (2010) also proposed the hbar-index, that is to say if the citation frequency of the paper is lower than the h-index of the scholar's partner, the paper will not be included in the literature collection of hbar-index.

As for the shortcomings that the h-index values of some scholars are common, the h_m-index proposed by Zhang (2007) can effectively solve this problem. Here is the calculation formula: $h_m = h + h/The~author's~cumulative~citation~frequency$. h_m-index can distinguish scholars with same h-index, and consider all the paper citation of a scholar, which is a better evaluation of scholars (Long et al. 2017).

On the question that h-index cannot evaluate scholars' short term contributions and it's value keeps rising, the h_{1_n}-index and the h_n-index were put forward by Yu and Wang (2017). The h_{1_n}-index refers to the h-index of a periodical in recent n years (n is the peak cited year for journal papers), and the h_n-index refers to the h index when the journal paper reaches the maximum cited peak. Furthermore, Lv et al. (2017) conducted an empirical analysis of four scholars through the h-index growth curve, and the following two conclusions are drawn: One is that h-index growth curve is more comprehensive, and the other one is that large scientific research team has the phenomenon of collaborative growth of h-index.

According to the current research, the research on h-index is mainly aimed at solving the problems that h-index is not sensitive to highly cited papers, and it does not consider the level of the partner and the time factor. The emergence of h-type index based on h-index has enriched the h-index (Yu and Wang 2017). But h-index and h-type index lack comprehensive evaluation of scientists' overall academic contribution.

3 Methods

Weighted h-index model, which is based on the h-index, calculates weighted composite scores according to the cited papers' frequency. It is a comprehensive evaluation index, not only manifesting the highly cited papers, but also revealing the value of the lowly cited paper. The weighted h-index includes h_w-index and h_{w_t}-index. h_w-index evaluates the whole academic career of scholars, and h_{w_t}-index lays particular stress on scholars' recent performance.

h_w-index is defined as: Assuming that a scholar has published N papers, and L is the result of summing the cited frequency of h academic papers that the cited frequency is higher than the others. Then calculate each paper's weights named C_n/L, and use C_n which is the each cited paper's frequency to multiply C_n/L. Finally accumulate the results of $C_n * C_n/L$ and use the cumulative results to multiply h-index to get the h_w-index. The following calculation formula is h_w-index.

$$h_w = ((C_1/L) * C_1 + (C_2/L) * C_2 + \ldots\ldots + (C_n/L) * C_n) * h \qquad (1)$$

simplify Formula (1):

$$h_w = (\Sigma C_n^2) * (h/L) \quad (n = 1, 2, 3 \ldots\ldots) \qquad (2)$$

Among them, C_n^2 is the cited paper's frequency square, and h is the h-index of scholars; L is the cumulative results of the cited frequency of h academic papers which the cited frequency is higher than the others. C_n^2 not only widens the gap between highly cited papers and other papers, but also highlights the value of highly cited papers, and lets the h-index sensitive to highly cited papers. But the disadvantage of the C_n^2 is that some highly cited papers effect too much. For example, the C_n^2 value of one paper cited 1000 times is better than 10 papers which is cited 316 times. Therefore, we use h-index to multiply C_n^2/L, in order to weaken the influence of some extremely highly cited papers, and at the same time, it reveals the importance of core paper.

However, the h_w-index cannot be used to do short-term evaluations, which cannot reflect the scholar's recent academic contribution and it is unfair to young scholars. Thus, based on h_w-index, we construct the h_{w_t}-index, giving different weights for papers published in different years. In order to simplify the formula as well as calculation, and increase the usability of the calculation model, we give p weight for papers that published nearly five years, and others are given $1 - p$ weight. So we get the h_{w_t}-index as formula (3).

$$h_{w\_t} = \left(\Sigma C_{t \leq 5}^2 * p + \Sigma C_{t > 5}^2 * (1 - p)\right) * (h/L) \quad (p \in [0,1], \quad p > 1 - p) \quad (3)$$

In formula (3), t represents when the papers were published, and $C_{t \leq 5}$ means the citation frequency of the paper published in nearly 5 years, while $C_{t > 5}$ means the citation frequency of the paper published more than 5 years. Besides, P is the weight of the paper contributing to h_{w_t}-index. h_{w_t}-index not only focuses on the academic performance of scholars in the past five years, but also considers the overall academic level of scholars. Comparative analysis of h_w-index and h_{w_t}-index can effectively analyze the comprehensive quality of scholars and the development status of scientific research in recent years.

4   Empirical Analysis

4.1   Literature Data Analysis

In order to verify the effectiveness of the model, 30 scholars in Library and Information Science field were selected for analysis. The literatures data of scholars are derived from Chinese citations database, and the retrieval time is November 5, 2017.

In Table 1, some scholars have the same h-index and g-index, and it is indistinguishable from the same value. Considering the cited frequency of h academic papers that the cited frequency is higher than the others, e-index solves the problem of same values, and the 30 scholars that we selected have different g-index.

Table 1. h-index, g-index and e-index of 30 scholars.

Scholar	h-index	Rank of h-index	g-index	e-index	Scholar	h-index	Rank of h-index	g-index	e-index
Qiu Jun-Ping	54	1	82	51.25	Xu Xin	22	16	43	32.68
Ke Ping	33	2	48	29.39	Si Li	21	17	33	21.17
Zhu Qing-Hua	31	3	51	34.70	Xie Yang-Qun	21	17	34	23.79
Su Xin-Ning	30	4	47	31.42	Li Gang	20	19	29	17.58
Huang Lu-Cheng	29	5	61	47.11	Chu Jie-Wang	19	20	32	21.68
Xiao Xi-Ming	28	6	43	27.59	Yuan Qin-Jian	18	21	26	15.46
Ma Hai-Qun	27	7	44	28.28	Deng Zhong-Hua	17	22	30	22.25
Liu Zi-Heng	25	8	44	31.64	Lu Zhang-Ping	16	23	24	15.20
Zheng Jian-Ming	24	9	45	33.42	Chen Fu-Ji	15	24	22	12.96
Huang Ru-Hua	24	9	35	20.69	Tang Xiao-Bo	15	24	23	14.42
Bi Qiang	24	9	33	18.06	Lan Yue-Xin	14	26	26	19.36
Lou Ce-Qun	24	9	39	26.36	Chen Cheng	14	26	20	11.31
Zhao Rong-Ying	23	13	39	27.13	Ma Xiao-Ting	14	26	19	10.91
Sun Jian-Jun	23	13	40	28.25	Zhang Xiang-Xian	12	29	21	14.49
Sheng Xiao-Ping	23	13	37	24.72	Deng Fu-Quan	9	30	14	9.33

The Table 2 shows the difference between the highest score of h_w and the lowest score is 9967, and the gap is large. Compared with the e-index, the h_w-index has expanded the distinction between scholars and considered the quality of the previous h papers, which is a comprehensive evaluation for scientific researchers. Compared with h-index, h_w-index has solved the problem of same values. Since it has considered scholars' whole papers, h_w-index is more scientific and reasonable than h-index.

Table 2. The results of h_w-index

Scholar	h-index	Rank of h-index	L	ΣC_n^2	h_w-index	Rank of h_w-index
Qiu Jun-Ping	54	1	5543	1052193	10250	1
Huang Lu-Cheng	29	5	3060	920815	8727	2
Zhu Qing-Hua	31	3	2165	266918	3822	3
Ma Hai-Qun	27	7	1529	215617	3807	4
Zheng Jian-Ming	24	9	1693	207679	2944	5
Xu Xin	22	16	1552	200264	2839	6
Xu Xin-Ning	30	4	1887	176511	2806	7
Ke Ping	33	2	1953	164988	2788	8
Liu Zi-Heng	25	8	1626	175132	2693	9
Sun Jian-Jun	23	13	1327	144068	2497	10
Xiao Xi-Ming	28	6	1545	131262	2379	11
Lou Ce-Qun	24	9	1271	113042	2135	12
Zhao Rong-Ying	23	13	1265	105671	1921	13
Deng Zhong-Hua	17	22	784	86977	1886	14
Xie Yang-Qun	21	17	1007	90338	1884	15
Sheng Xiao-Ping	23	13	1140	84504	1705	16
Bi Qiang	24	9	902	63935	1701	17
Chu Jie-Wang	19	20	831	59899	1370	18
Huang Ru-Hua	24	9	1004	55471	1326	19
Si Li	21	17	889	53965	1275	20
Li Gang	20	19	709	40681	1148	21
Lan Yue-Xin	14	26	571	39024	957	22
Yuan Qin-Jian	18	21	563	28578	914	23
Lu Zhang-Ping	16	23	487	27338	898	24
Tang Xiao-Bo	15	25	433	17868	619	25
Chen Fu-Ji	15	24	393	16106	615	26
Ma Xiao-Ting	14	26	315	12047	535	27
Chen Cheng	14	26	324	11785	509	28
Zhang Xiang-Xian	12	29	354	13604	461	29
Deng Fu-Quan	9	30	168	5282	283	30

Compared the rank of h_w-index with the rank of him-cited h_w-index, we know that self-citation raised the scholar's score, and higher self-cited scholars have higher scores than the same level of scholars who are less self- cited. Despite the change in ranks, there is little difference in their scores between scholars whose ranking has changed, generally within 50 points. Because ΣC_n^2 enlarges the influence of the cited frequency and the difference of scores between scholars, it is considered that the academic level among scholars is similar or even identical when there is not much difference in scores. We think the h_w-index weakens the effect of self-citation on the final score Table 3.

Table 3. h_w-index and him-cited h_w-index

Scholar	h-index	h_w-index	Rank of h_w-index	him-cited h-index	him-cited h_w-index	Rank of him-cited h_w-index
Qiu Jun-Ping	54	10250	1	51	9719	1
Huang Lu-Cheng	29	8727	2	28	8440	2
Ma Hai-Qun	27	3807	4	27	3703	3
Zhu Qing-Hua	31	3822	3	30	3683	4
Zheng Jian-Ming	24	2944	5	24	2927	5
Xu Xin	22	2839	6	22	2830	6
Ke Ping	33	2788	8	33	2721	7
Su Xin-Ning	30	2806	7	29	2701	8
Liu Zi-Heng	25	2693	9	24	2569	9
Sun Jian-Jun	23	2497	10	22	2387	10
Xiao Xi-Ming	28	2379	11	27	2288	11
Lou Ce-Qun	24	2135	12	23	1936	12
Xie Yang-Qun	21	1884	15	21	1867	13
Zhao Rong-Ying	23	1921	13	23	1832	14
Deng Zhong-Hua	17	1886	14	16	1818	15
Sheng Xiao-Ping	23	1705	16	23	1671	16
Bi Qiang	24	1701	17	24	1638	17
Chu Jie-Wang	19	1370	18	19	1344	18
Huang Ru-Hua	24	1326	19	24	1302	19
Si Li	21	1275	20	20	1217	20
Li Gang	20	1148	21	20	1112	21
Yuan Qin-Jian	18	914	23	18	890	22
Lu Zhang-Ping	16	898	24	15	837	23
Lan Yue-Xin	14	957	22	12	742	24
Tang Xiao-Bo	15	619	25	15	607	25
Chen Fu-Ji	15	615	26	14	577	26
Ma Xiao-Ting	14	535	27	13	504	27
Chen Cheng	14	509	28	14	490	28
Zhang Xiang-Xian	12	461	29	12	426	29
Deng Fu-Quan	9	283	30	9	261	30

In order to verify whether the formula (3) can better reflect the research work and the contribution of scholars in recent years, and weaken the negative influence of h_w-index on young scholars, we introduce the age information of scholars to help us analyze. The literature published since 2012 (about 5 years) are given the weight of 80 percent in this paper while others are given the weight of 20%. In practical application situation, the proportion of enactment can be set according to the needs.

After giving different weights, the rank of h_{w_t}-index of scholars generally change comparing with h_w-index. The change of Zhao Yu-Xiang borned in 1983 is most obvious, and his h_w-index ranks 15 while h_{w_t}-index ranks 6. His total citation frequency is not high, but his scientific research work in recent years is excellent. Thus, the h_{w_t}-index highlighting the contributions of scholars in recent years, while it has not ignored the contributions of scholars in the past Table 4.

Table 4. h_w-index and h_{w_t}-index

Scholar	Age	Rank of age	h-index	The quoted sum of squares since 2012	h_w-index	Rank of h_w-index	h_{w_t}-index	Rank of h_{w_t}-idex
Qiu Jun-Ping	1947	1	54	40483	10250	1	2287	1
Huang Lu-Cheng	1956	5	29	6636	8727	2	1783	2
Zhu Qing-Hua	1963	15	31	44828	3822	3	1150	3
Ma Hai-Qun	1964	16	27	6829	3807	4	834	4
Su Xin-Ning	1955	4	30	12277	2806	7	678	5
Zhao Xiang-Yu	1983	23	14	30499	1190	15	664	6
Xu Xin	1976	22	22	9942	2839	6	652	7
Ke Ping	1962	11	33	7187	2788	8	630	8
Zheng Jian-Ming	1960	8	24	3970	2944	5	623	9
Sun Jian-Jun	1962	11	23	8612	2497	9	589	10
Lou Ce-Qun	1956	5	24	4094	2135	10	473	11
Huang Ru-Hua	1968	18	24	13480	1326	14	459	12
Bi Qiang	1954	2	24	5616	1701	12	430	13
Xie Yang-Qun	1962	11	21	3133	1884	11	416	14
Li Gang	1966	17	20	8618	1148	16	375	15
Yuan Qin-Jian	1969	19	18	9766	914	17	370	16
Lu Zhang-Ping	1958	7	16	8821	898	18	354	17
Chu Jie-Wang	1969	19	19	4242	1370	13	332	18
Chen Cheng	1974	21	14	8526	509	21	323	19
Tang Xiao-Bo	1962	11	15	8390	619	19	298	20
Chen Fu-Ji	1954	2	15	6897	615	20	281	21
Zhang Xiang-Xian	1960	8	12	3767	461	22	169	22
Deng Fu-Quan	1961	10	9	214	283	23	63	23

4.2   Results and Discussions

Based on empirical analysis of h_w-index and h_{w_t}-index, we know that they have following characteristics:

(1) The h_w-index can highlight the value of the highly cited papers, and it is a comprehensive evaluation index considering the contribution of scholar's each paper. Based on L (the cumulative citation frequency of the previous h papers), construct the weight of C_n/L, and then times h-index. h_w-index shows the contribution of the overall high citation papers of scholars. C_n^2 expands the influence of highly cited papers, especially the contributions of extremely high cited papers. In the formula (2), the citation frequency of each paper is used, and the final score results are influenced by all literatures.

(2) Compared with h-index, h_w-index is easier to distinguish scholars' contributions. According to the results of h_w-index calculated by 30 selected scholars, we are easy to know that the h_w-index of the scholars does not have the same score, which solves the problem that some scholars' h-index values are common.

(3) The h_w-index weakens the influence of self-citation on evaluation. Obtaining h-index, him-cited h-index, h_w-index and him-cited h_w-index from Table 3, we know that the score of h_w-index was almost identical to him-cited h_w-index. Therefore, the h_w-index weakens the effect of self-citation on the results.

(4) h_{w_t}-index highlights the value of scientific research achievements in recent years, and also considers the scholars' past contributions. In the short-term evaluation, it is more reasonable for young scholars and scholars who have made great contributions in the past. After giving different weights, the scholars who have published more papers in recent years have a higher score than other scholars.

(5) Comparing h_w-index with h_{w_t}-index, potential scholars can be found. The h_w-index and h_{w_t}-index of researchers in the same research field were calculated and the two indexes were sorted respectively, so we will discover potential researchers by comparing the changes of the scores of two indexes and analyzing the scientific research of scientists in recent years. It can be used for the excavation and recruitment of talents.

5   Conclusion

Considering the contribution of scientists' previous h academic papers, this paper proposed two weighted h-index models which are named h_w-index and h_{w_t}-index, and then selected 30 active Chinese scholars in recent years in Library and Information Science field for empirical analysis. Revealing highly cited papers and considering the contribution of scientists' whole papers, h_w-index not only weakens the influence of self-citation on the results, but also makes it easy to distinguish scientists' contributions. The h_{w_t}-index focuses on the research output of scientists in recent years. Meantime, it also considers their past contributions. Therefore, in short term evaluation, the h_{w_t}-index is more reasonable for young scientists and scholars who have made a great contribution in the past years. Potential scholars can be identified by the way of comparing h_w-index and h_{w_t}-index.

Due to the length of space, this paper has not discussed the optimal weight ratio of h_{w_t}-index and the division of score interval of h_w-index and h_{w_t}-index. In follow-up studies, we will consider the factors of scholars' partner, further exploring the optimal h_{w_t}-index weight and the division of the score interval to increase the operability in the practical evaluation of h_w-index and h_{w_t}-index.

Acknowledgements. Publication of this article was funded by National Social Science Foundation of China (Grant numbers 16BTQ071).

References

Hirsch, J.E.: An index to quantify an individual's scientific research output. Proc. Natl. Acad. Sci. U.S.A. **102**(46), 16569–16572 (2005)

Rodrigo, C.: The h-index: advantages, limitations and its relation with other bibliometric indicators at the micro level. J. Informetr. **1**(3), 193–203 (2007)

Egghe, L.: Theory and practice of the g-index. Scientometrics **69**, 131–135 (2006)

Zhang, C.: The e-index, complementing the h-index for excess citations. PLoS One **4**(5), 1–4 (2009)

Wu, Q.: The w-index: a measure to access scientific impact by focusing on widely cited papers. J. Am. Soc. Inform. Sci. Technol. **61**(3), 609–614 (2010)

Hirsch, J.E.: An index to quantify an individual's scientific research output that takes into account the effect of multiple coauthorship. Scientometrics **85**(3), 741–754 (2010)

Sui, G.: Relations among g, h and e indexes and its bibliometrics meaning. J. Libr. Inf. Serv. **57**(23), 90–94 (2013)

Yu, L., Wang, Z.: A perspective of time h index innovatoin: h1_n index and hn index. J. China Soc. Sci. Tech. Inf. **36**(04), 346–351 (2017)

Zhang, X.: hm Index-A modification to h index. J. Libr. Inf. Serv. (10), 116–118+16 (2007)

Long, Y., Zhao, Q., Zhao, X.: Q-index—A new type of h-index considering time. J. Libr. Inf. Serv. **61**(07), 91–95 (2017)

Lv, N., Liu, Y., Quan, S.: h-index time trends analysis based on author contribution. J. Intell. **34**(04), 54–58 (2017)

Decision Analysis Based on Artificial Neural Network for Feeding an Industrial Refrigeration System Through the Use of Photovoltaic Energy

Fabio Porreca[(⊠)]

University of L'Aquila, L'Aquila, Italy
fabioporreca@enertecnology.com

Abstract. The evaluation of the energy availability from renewable sources in the industrial processes is at the basis of many researches in engineering. The non-programmable nature of many of these sources often leads to consider them as a simple support and not as a primary source of supply. With this in mind, this research has been directed to try to exploit the forecasting abilities of the neural networks in order to create scenarios applicable in different high-energy consuming industrial contexts which reckon the optimization of the energy consumption as the new objective of the so called "green business".

Keywords: Artificial neural network · Data mining · Renewable energy

1 Introduction

After years of planning in the renewable energies field, for plants often dictated more by the economic desirability of the government incentive than by the proper objective of optimising the company consumption, the intention to extend the analysis to a wider spectrum has led to new synergies and opportunities for innovative business. Energy must not be seen as a burden to bear (consumption) but as an opportunity to create new models of investments. The particular example of self-produced renewable energies applied to high-energy consuming processes is further emphasized when the production curve matches perfectly the necessity one in order to minimize costs and shorten the time for the investment return [1].

The industrial processes of this typology are characterized by absorption profiles that, after an initial transitional period, defined by peaks of energy request, stabilise on a constant or predictable consumption level enabling to assess the needs and to manage their significant impact on the energy budget and, consequently, on the economic one. Interesting examples are those implying the controlled heat generation and/or dissipation in materials. The latter, the controlled cooling, has given an interesting input in the management of the controllability of the process matched with the prediction of the energy production from photovoltaic sources.

There are many researches aimed at making the production of renewable energies predictable but, in most of the cases, they are directed to the need to evaluate the

© Springer Nature Switzerland AG 2019
E. Bucciarelli et al. (Eds.): DCAI 2018, AISC 805, pp. 134–142, 2019.
https://doi.org/10.1007/978-3-319-99698-1_15

quantity of exportable energy to the distribution networks in order to better calibrate the "energy mix" of the power exchange [2]. In the present analysis, the aim instead is to determine also the reachable peak powers and how long they can be available during the day, in order to properly assess "the tracking of the production curve".

2 Analysis Methodology Adopted: CRISP-DM (CRoss Industry Standard Process for Data Mining)

In the evaluation of the problem it has been adopted a method widely used in industry: the CRISP-DM – CROss Industry Standard Process for Data Mining [3].

It is a cyclical process based on six steps:

- **Business Understanding:** establishing the objective of the research aimed at creating and optimising the business project and at directing its development.
- **Data Understanding:** evaluation of the data to collect in order to achieve results, their availability, difficulties in finding them and possible aggregation processes.
- **Data Preparation:** preparation and collection of the data identified, their selection, elimination of redundant and unnecessary information, draft.
- **Modeling:** data analysis process through mathematical instruments that allow the extrapolation of the expected results and the creation of models by means of the subdivision into calculation and verification groups.
- **Evaluation:** evaluation of the results of the implemented model in the contest of the business objective analysed through various iterations of the aggregation, statistical and artificial intelligence instruments, that allow the correlation of the various sets of data.
- **Deployment:** the data collection is thus used to verify the preliminary assumptions made and identify unexpected results. Through the knowledge acquired it is possible to address the technical-economical choices and, possibly, to modify the parameters chosen and follow the process towards further results.

2.1 Business Understanding

The industrial cooling process, regardless of the carrier used (Freon, Ammonia or Carbon dioxide) rests essentially on the refrigeration cycle based on the compression and subsequent expansion of the fluid; the resulting change of state requires a lot of thermal energy that is absorbed from the environment in which the fluid is led, in systems equipped with fans called heat exchangers. All the system is equipped with electric pumps, expansion tanks and, first of all, with compressors powering the cycle by absorbing big quantities of electric energy.

This electric energy consumption is the most important cost element in the cooling process and its optimization, even if slightly, can make the difference in the freezing business characterized by large volumes and relatively low prices when compared to the final value of the product.

Connecting the use of renewable sources to such an energy consuming process can create a virtuous cycle and excellent economic returns. In particular, considering the photovoltaic energy leads to significant results thanks to the production peak that these systems have in the summer period when the external temperatures are higher, and thus the request of refrigerating power for the freezing is higher as well.

The limit of this renewable energy power, as all the others, is the insufficient, if not absent, possibility to know in advance the quantity and the production level they will have, thus enabling to match them to the number and the switching times of the machineries. With this in mind the aim is to try to establish a criterion that allows estimating sufficiently in advance the energy production and try to store it in order to control its availability basing on company requirements. Considered the significant dimension of the power at stake, hundreds of kilowatts, and the necessary energy, the use of electrochemical storage is not an option because of the high costs.

This necessity has found a solution with the idea of a thermodynamic battery based on the structure of the refrigeration system itself. Using a concentric architectural system in which the colder part is put inside and then gradually the "warmer" ones, it is possible to minimize the dispersion and, at the same time, to reuse it to limit the needs of the outer layers. These latter, being at a relatively higher temperature, have less difference in degree with the environment and so require less energy to compensate its thermal variations.

Thus, the possibility to evaluate the energy availability allows deciding when and for how long to use the refrigerating machines and when, instead, to use the frigories stored inside in the "cold" part of the building.

2.2 Data Understanding

The productivity analysis of a photovoltaic system can be done considering the different associated conditions characterizing the installation site and the possible evolutions that the weather conditions can present in a certain period. The study deals with the prediction of the power that can be generated by a photovoltaic plant across a variable time period (max + 24 h) [4].

The data to be used in the evaluation have been collected by a site kept under continuous observation with a SCADA system recording on DB both the data coming from the meteorological station (environment temperature and irradiation) and the ones of the production system through sensors (cell temperature) and bus RS485 on the inverters (power).

The set of the extracted data consists of lists with 4 different parameters taken every 10 min for a year in order to cover the different weather conditions that may arise. The measures have been exported to a spreadsheet in order to be selected and processed later.

2.3 Data Preparation

The site chosen for the survey is located in centre Italy, in the province of Chieti, in the municipality of Atessa.

The production has been recorded by the inverters on a plant of about 600 kWp, placed on a homogeneous SOUTH – facing cover with 10° tilt.

The basis of the data is given by readings taken every 10 min over a time period of a solar year.

The extrapolation has involved the following parameters:

- External temperature
- Temperature of the photovoltaic cell taken on the module back sheet
- Irradiation intensity on the modules plane
- Power generated by the plant

For a better network response, it has been decided to split the data at disposal into two seasonal groups and to reduce the measures sampling from 10 to 30 min:

- SEASON 1: period NOVEMBER–APRIL
- SEASON 2: period MAY–NOVEMBER

2.4 Modeling

Our problem appears as a time-discrete, univariate dynamic system and with more input variables for which it has been used a particular neural network (ANN – Artificial Neural Network) characterized by a MLP approach (Multi- Layer Perceptron) and, in particular, a non-linear autoregressive ANN with exogenous input [5] (Fig. 1):

$$y(t) = f(y(t-1), (y(t-2), \ldots, y(t-d), x(t-1), x(t-2), \ldots, x(t-k)) \qquad (1)$$

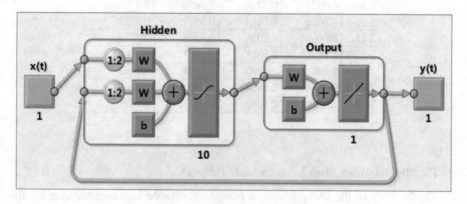

Fig. 1. Neural network scheme

Exogenous input x(t): external temperature (TEMP_EST) and solar irradiance (IRR)

Recursive input y(t-d): time series of the predicted power
Output y(t): predicted power in the subsequent hours

2.5 Evaluation

Below an example of evaluation made on SEASON 1.

The following indicators have been used as errors to evaluate the performance of the network:

- nRMSE – Normalized Root Mean Square Error

$$nRMSE = \frac{1}{Ymax - Ymin} \sqrt{\frac{\sum_{t=1}^{n} (\hat{y}_t - y_t)^2}{n}} \tag{2}$$

cv_RMSE – Coefficient of Variation RMSE

$$cv_{RMSE} = \frac{RMSE}{\bar{y}} \tag{3}$$

Table 1. Evaluation of errors

ANN	n_RMSE_t	n_RMSE_d	cv_RMSE_t	cv_RMSE_d
2-5-1	0.0667	0.1021	0.5620	0.7734
2-6-1	0.0660	0.1036	0.5564	0.7841
2-7-1	0.0683	0.1062	0.5758	0.8044
2-8-1	0.0673	0.1034	0.5676	0.7832
2-9-1	0.0671	0.1028	0.5658	0.7782
2-10-1	0.0666	0.1030	0.5619	0.7797
2-11-1	0.0660	0.1026	0.5568	0.7772
2-12-1	0.0651	0.1018	0.5487	0.7707
2-13-1	0.0687	0.0990	0.5795	0.7498
2-14-1	0.0660	0.1034	0.5562	0.7829
2-15-1	0.0650	0.1040	0.5477	0.7876

- Thus the following chart has come out (Table 1):

On the basis of the above results, it comes out that the best configuration for the neural network results to be 2-13-1 – 2 input neurons, 13 neurons for the hidden layer and 1 output neuron.

In order to make the result of the simulation more comprehensible it is possible to notice, from the attached graphs, the good overlapping between the real data and the predicted ones in the training and in the testing phase (Figs. 2, 3, 4 and 5):

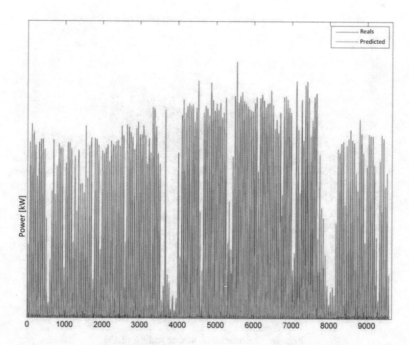

Fig. 2. Predicted-real power training phase

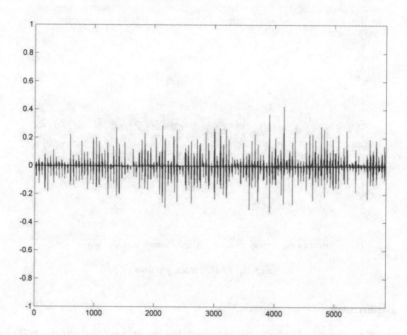

Fig. 3. RMSE training phase

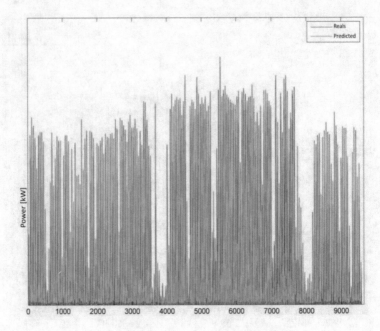

Fig. 4. Predicted-real power testing phase

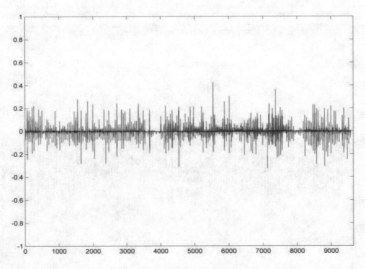

Fig. 5. RMSE testing phase

2.6 Deployment

The example given clearly shows the potential that the neural networks can offer to the prediction, within the following 24 h, of the available power of the photovoltaic source. It was deliberately decided to present the result of the SEASON 1 because they are the ones interested by less available power and higher variability due to the meteorological changes and thus it shows even better the utility of this data mining tool in order to improve the managing information of the energy consuming site.

Subsequently to these results it has been created a system based on PLC that evaluates in real-time the energy availability and turns consequently on the engines to optimize the consumption of the plant (Fig. 6):

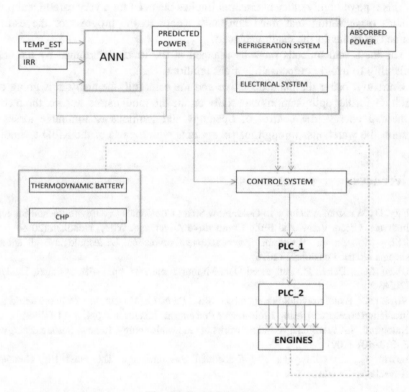

Fig. 6. Control system

3 Conclusions and Future Research

The analysis, based on real data and connected to a real problem, has led to the creation of a logistic hub for the goods conservation in controlled temperature from +9 ° a −35 °C characterized by an excellent energy performance with a consequent reduction of the average cost of the service by approximately 30%.

This forecasting has also been followed by an implementation strongly focused on energy saving and heat dispersion minimization, which has further contributed to reduce consumption and energy requirements. The green economy philosophy, linked to an innovative engineering approach, can convincingly demonstrate how companies can better exploit global challenges to achieve long-lasting profits and growth.

Companies like these are pursuing a profit change in the traditional markets.

Besides offering smarter (rather than just greener) products for consumers in general, the integrated sustainability can greatly motivate employees. Most of all, it enables companies to generate even higher returns for investors, while responding to new market realities in consideration of the declining energy and technology resources, of the processes transparency and the growing customers' expectations about convenience and environment sustainability [6].

This type of highly efficient creation implies the need for a very careful and precise logistics organization that must constantly adapt to the progress of the estimates obtained to maintain the result achieved.

The loads and unloads must be planned at the times determined by the energy availability to avoid compensating night ignitions.

Currently, being the plant operative and the cells full, the analysis is going on in search for further optimizations especially on the thermodynamic storage, the recovery of thermal energy, the control of openings and ignitions to minimize losses and organize the warehouse through matrix systems with the use of the RFID technology.

References

1. Esty, D., Winston, A.: Green to Gold: How Smart Companies Use Environmental Strategy to Innovate, Create Value, and Build Competitive Advantage. Wiley, Hoboken (2009)
2. RSE - Monogra-fia - I sistemi di generazione fotovoltaica: La tecnologia e gli effetti sul sistema elettrico nazionale (2016)
3. Olson, D.L., Delen, D.: Advanced Data Mining Tecniques, pp. 9–19. Springer, Heidelberg (2008)
4. Voyant, C., Randimbivololona, P., Nivet, M.L., Paoli, C., Muselli, M.: 24-hours ahead global irradiation forecasting using Multi-Layer Perceptron. Meteorol. Appl. 6–11 (1999)
5. Kalogirou, S.A.: Artificial neural networks in renewable energy. Renew. Sustain. Energy Rev. 5, 373–401 (2001)
6. Laszlo, C., Zhexembayeva, N.: Embedded Sustainability: The Next Big Competitive Advantage, 1st edn

Exit, Voice and Loyalty in Consumers' Online-Posting Behavior: An Empirical Analysis of Reviews and Ratings Found on Amazon.com

Tony E. Persico[1]([✉]), Giovanna Sedda[2], and Assia Liberatore[1]

[1] Research Centre for Evaluation and Socio-Economic Development,
University of Chieti-Pescara, Chieti, Pescara, Italy
to.persico@gmail.com, assia.liberatore@unich.it
[2] B-Eat Digital Agency, Rome, Italy

Abstract. In this paper, we aim to describe the behavior of e-commerce consumers by analyzing the distribution of online reviews and ratings. Different from studies conducted previously, which have focused on the positivity and negativity of ratings, our work analyzes the ratings distribution through a tensor-based approach. This approach allows us to observe a new range of information related to distributions' features that we describe through the "Exit, Voice and Loyalty" scheme. In addition, we seek a distribution function capable of capturing these features, and we aim to over-perform the synthesis provided by using a polynomial regression model. For this reason, we introduce an *ad hoc* beta-type modified function to create a proxy of collected data. We found a tri-modal distribution (S-modal) as a relevant component of the J-shaped distributions referred in the literature.

Keywords: E-commerce · Product reviews · Search and experience goods
Consumer behaviour · Big data

1 Introduction

The increase of online commerce has created a new channel of communication among customers, as well as opening dialogue between customers and producers. The review system, in a social context, has prompted exponential growth of word-of-mouth communication [1] where consumers can express more clearly what they desire, and may verbalize the intensity of their dissatisfaction more easily. Reviews have given birth to social thought and have provided an extraordinary source of information for the decisions theory [2]. Following [3], we interpreted consumer exit, voice and loyalty as strong indicators of rising consumer power on the Internet. Recently, economists began studying the impact of this new element to consumers choices in conjunction with other disciplines, such as psychology and cognitive sciences (among other see [4] for the framing effect and [5–7] for exploratory behavior). Reviews impact both the sale of the products and willingness to pay [8, 9]. Current literature on this issue focuses on the disparity between the effects impacts of positive and negative reviews [10, 11]. Additionally, reviews have been examined in the context of their helpfulness,

© Springer Nature Switzerland AG 2019
E. Bucciarelli et al. (Eds.): DCAI 2018, AISC 805, pp. 143–153, 2019.
https://doi.org/10.1007/978-3-319-99698-1_16

effectiveness [12, 13] and order of display – from both a qualitative and quantitative point of view [14]. Analysis has also been conducted on self-selection bias [15, 16]. Researchers describe the distribution of ratings as a deviation of normal or unimodal distributions, while quantitative analysis on the informative role of distributions is limited [17]. In current literature, we found two main gaps, which we have tried to fill in this research. First, on the data synthesis side, is the distinction a priori between negative and positive feedback and connected behaviors from the consumers. This distinction has not allowed for a deeper study of behavior defined as the full group of actions that lead to a review (purchase, decision to write, type of information provided). The second, then, is the lack of analysis on aggregated behaviors. Thus, we want to address these two main points by first studying the full array of information carried by reviews, specifically by their associated ratings, and examining specific behaviors (i.e., decisions) more complex than the common negative/positive labeling. In addition, second, we aim to study how these specific decisions can be aggregated and to analyze the features of these social behaviors.

Our research aims to consider each set of reviews for a product as a single object of analysis, i.e., a vector of frequency of related ratings. This is possible following the tensor-based approach [18], which recognizes the features of each distribution, introducing an unexplored path of study. According to [14], consumers have the possibility to post product reviews with content in the form of numerical star ratings and in the form of open-ended customer-authored comments about the product. Considering the two forms mentioned above, we focused on star ratings because their numerical form can be easily statistically managed. If you are not familiar with online reviews, think of them as normally distributed. Experience suggests that the distribution of the ratings follows a decreasing trend, from the highest to the lowest. Some researchers describe this trend in an intuitive way, as J-shaped, referring to the low frequency of intermediate ratings [15]. We analytically investigate this type of distribution to find recurring elements and a descriptive synthetic function. Our goal is to highlight main features in the distributions of ratings and to describe them through a synthesized function. In Sect. 2, we describe our sampling method and its composition based on online ratings among different categories of products. In Sect. 3, we underline the composition of typical distributions for each type of product, as well as introduce our ad hoc function of distribution. In Sect. 4, we discuss our main results, which reveal the prominence of tri-modal distributions. In Sect. 5, we place our overall results in the context of current literature and highlight directions for further research.

2   Dataset Collection and Method

For our study, we selected 1665 products on sale on amazon.com. This website was chosen because of its leading role in e-commerce and the depth of available reviews [19]. In addition, it is commonly used by scholars for data collection [20]. We examined approximately half a million reviews (exactly 503.237 single ratings) collected in the period between September 10, 2017 and September 15, 2017. Because data collection occurred on an American website, the dates were chosen to avoid possible seasonal deals (Black Friday, winter holiday season, etc.). The quality of the sample is consistent with that of other research samples both in term of products and reviews [21, 22]. However, there are examples wherein researchers used larger [12–14] or smaller samples [15]. In general, we find the quality consistent with the composition of a sufficiently balanced dataset. Nelson [23] divided products into search products and experience products. We selected categories including both "search" products, for which consumers can evaluate quantitative and objective information, and a category of "experience" products (clothes and accessories), for which evaluation depends on both qualitative and subjective information [24, 25]. These categories are as follows: (1) accessories; (2) consumption (3) complex products; (4) clothes; (5) grocery. In line with [25], total price paid, product familiarity and Internet shopping experience may affect consumer searches, as well as socio-demographic variables such as household income, highest education level for any member of the household, age of the eldest head of household, household size, and the presence of children. We built categories basing on total price paid. Below, we show the composition of categories based on total price paid. In the first category, we gather accessories and instruments that can absolve only one function (e.g., phone holder, flashlight) and that cost below $30. The second category includes consumer goods used daily (e.g., batteries, pens) with a price under $30. The complex products category consists of items designed for multiple functions (e.g., notebook, smartphone, all-in-one printer) that cost over $100. The grocery products category includes products that are consumed daily that cost below $30. In the clothes category, we separated men's and women's clothing and accessories that cost over $30. The first four categories can be included in the "search" group of products, as our main focus is on this type of product. The last category can be referred to the "experience" group and is utilized as a control group. To begin, we attempted to distinguish between men's and women's clothing, as it is often found in the literature (the lateral difference was observed in [5, 6]. This distinction did not reveal statistically significant differences, so we elected to exclude it. Additionally, we noted that even if grocery products showed overlapping characteristics among the search and experience products, they registered results similar to "search" products (probably due to the high standardization of the articles on sale).

The statistical sample used herein can be defined as pseudo-random [26]: we picked objects that could be organized into discrete categories and that have more than 100 reviews, selected in random intervals (determined by random numbers from 1 to 10), starting with the first results of the search page. We chose this type of statistical sampling to obtain representative sample of the consumer buying experience (replicating the order of suggested items). As in [26], our dataset was constructed with an algorithm that contacts Amazon's public application programming interfaces (APIs) and collects ratings information for each product. We chose a minimum limit of reviews to avoid products with poor popularity as they are of little familiarity to users and thus would be unlikely to receive much attention. In contrast, we did not limit overly popular product items [14]. Figure 1 reports the synthesis of the observed distributions in the five product categories. For each category, we selected 333 products and registered the percentage composition of frequencies of their ratings. The resultant dataset of 1665 vectors, which sum up to the composition of the overall ratings collected, was analyzed using both qualitative and quantitative analysis. First, we described the common

characteristics among the frequency distributions found in the sample. Next, we identified a synthetic interpretation of data towards a specific function of distribution.

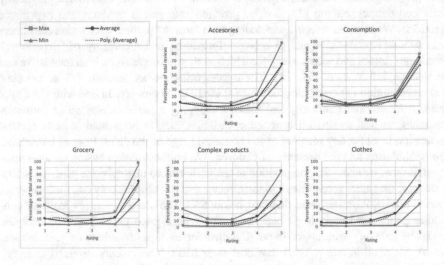

Fig. 1. Distribution of frequencies of products ratings per categories.

2.1 Qualitative Analyses

We considered the distribution of ratings for every item as a vector i of dimension 5
defined as x_{ir}: (x_{i1}, x_{i2}, x_{i3}, x_{i4}, x_{i5}), where each element reports the frequency of the
relative stars rating so that the whole dataset can be represented by a 1665×5 matrix,
where each category is a 333×5 sub-matrix. For each sub-matrix, we conducted a
qualitative analysis to identify the main features of the distributions utilizing the following filters:

Unimodal distribution: $x_{i1} \leq x_{i2} \leq x_{i3} \leq x_{i4} \leq x_{i5}$;

Bi-modal distribution: $x_{i1} > x_{i2}$ U $x_{i4} < x_{i5}$ U $x_{i3} < x_{i5}$; $x_{i1} < x_{i2}$ U $x_{i4} < x_{i5}$ U
$x_{i3} < x_{i5}$;

- U-modal distribution: $x_{i1} > x_{i2}$ U $x_{i4} < x_{i5}$ U $x_{i3} < x_{i5}$;
- S-modal: $x_{i1} > x_{i2}$ U $x_{i3} < x_{i4} < x_{i5}$;
- Strictly S-modal: $x_{i1} > x_{i2}$ U $x_{i3} < x_{i4} < x_{i5}$ U $x_{i1} < x_{i4}$;

We further differentiated bi-modal distributions – j-shaped – by identifying the sub
category U-modal distribution, and we added further differentiation to the U-modal to
identify the presence of an intermediate mode. In these cases, modes are represented by
the maximum rating (5), the rating immediately following (4) and by the minimum
rating (1). We defined S-modal as the vector where the frequency of the 4 star rating is
superior to the 3 and 2 stars rating. In addition, if the frequency of the 4 star rating is
superior to all the minor ratings (3, 2 stars and 1), we defined the distribution as strictly
S-modal. In the following section, we discuss the role of this last distinction. The
results of the qualitative analysis are reported in Table 1.

Table 1. Typologies of distribution of ratings among categories of products

Type	% EVL	Unimodal	Bimodal	U-modal	S-modal (Strictly S)
Consumption	93.2%	0.0%	100.0%	100.0%	100.0% (88.0%)
Complex product	87.4%	0.0%	100.0%	97.0%	97.0% (48.3%)
Accessories	88.2%	12.1%	87.9%	81.8%	81.7% (54.7%)
Grocery	87.5%	18.2%	81.8%	75.8%	57.7% (51.4%)
Search products	**89.1%**	**7.6%**	**92.4%**	**88.6%**	**84.1% (60.6%)**
Clothes (exp. products)	86.2%	63.6%	36.4%	36.4%	33.0% (30.3%)
Total	**88.9%**	**18.8%**	**81.2%**	**78.2%**	**73.9% (54.5%)**

2.2 Quantitative Analyses

To proceed with the analysis, we follow the tensor-based approach [18] consisting of three main steps: (1) the individuation of a distribution's function to describe the phenomena and the relative objective function; (2) the construction of the tensor with values estimated by the function; and (3) the estimation of the parameters of the function through the maximization of the objective function. We proposed two different distribution's functions and replicated for these steps for each. We found that the data collected can be described through histograms, an approximation of a discrete beta-type distribution. Because we encountered difficulties in identifying frequencies that were very high, the maximum rating, we introduced a new distribution function named P_{ab}, characterized by a parameter of exponential transformation, named n (i.e., the beta function to the power of n) where $n = 5$. The range of the parameter is essential to allow the function to capture a strictly U-modal distribution. The parameters a' and b' are independent and less than one. The function we used has the form:

$$f(\mathbf{x}) = P_{ab}(a', b', x_r) = \frac{\left[x^{a'-1}(1-x)^{b'-1} \right]^n}{\sum_{r=1}^{n} \left[x^{a'-1}(1-x)^{b'-1} \right]^n} \tag{1}$$

Under the following hypothesis:

(1) $n = 5$
(2) $1 > a' > b' > 0$
(3) $x \in [0,1]$
(4) $x = x_r/(n+1)$

In addition, we found that the values of the parameters were linked with the observed values of the frequency of the highest ranking. For this reason, we introduced an additional hypothesis to simplify the proposed estimation where the parameter b' is dependent from the parameter a. Then, (1) can be rewritten using only one parameter:

$$f(\mathbf{x}) = P_a(a, x_r) = \frac{[x^{a-1}(1-x)^{-a}]^n}{\sum_{r=1}^{n} [x^{a-1}(1-x)^{-a}]^n} \tag{2}$$

Under the hypothesis 1, 2, 3, 4 and the additional hypothesis:

(5) $b' = 1 - a$

as the (1) and the (2) are our distribution functions, we defined our objective function to maximize the correlation between the vector x_i and the vector obtained through the distribution function x_j:

$$F(x_i, f(x)) = corr(x_i, x_j) \tag{3}$$

In the second step, we created two tensors with all possible vectors x_i derived by P_{ab} and P_a. The first tensor of dimension $n \times m \times h$ was populated through the calculation of P_{ab} through two vectors: vector a with m values and vector b of h values populated with all possible values discrete of the respective parameters (with intervals of 0.01) comprised in the range of hypothesis 2. The second tensor can be reduced to a matrix of dimension $n \times m$ with the vector a following hypothesis 2 and 5. Next, we sought to identify parameters which maximize our objective function among product categories (as discussed in the following section, we focused on the product category). To perform this step, we calculated the correlation of each collected vector with each calculated vector to identify the maximum correlation within each product category as follows:

$$f(x) = \max F(x_i, f(x)) = \max \sum corr(x_i, x_j) \tag{4}$$

The maximization of the objective function with the greatest correlation the observed data allows us to estimate the related parameters for each product category. As we conducted the analyses both for the P_{ab} and P_a, we estimated the parameters for both distribution functions: a' b', and a. The results of this estimation are reported in Table 2.

Table 2. Correlation index between calculated and observed data and P function parameters estimations.

Category of products	(Part I) Correlation between distribution functions and observed data				(Part II) Parameters estimation for P_a and P_{ab} function		
	Poly 2	Poly 3	P_a	P_{ab}	a	a'	b'
Consumption	0.965	0.996	0.999	0.999	0.65	0.63	0.34
Complex product	0.983	0.996	0.996	0.997	0.58	0.65	0.48
Accessories	0.960	0.996	0.999	0.999	0.61	0.65	0.43
Grocery	0.950	0.990	0.999	0.999	0.62	0.61	0.37

3 Results

By analyzing the data from a qualitative point of view, we determined that the frequency of the ratings 5 stars, 4 stars and 1 star, indicated as exit, voice and loyalty (EVL), represent almost the 90% of all ratings assigned within the sample. From the application of filters, we derived the qualitative trend of each distribution, where the U-modal distribution is a particular case of the Bi-modal distribution and the S-modal distribution is a particular case of the U-modal distribution (and of course the strictly S-modal distribution is a particular S-modal distribution). All the frequency distributions have the highest rating as their mode, while more than the 80% of the distributions are at least bi-modal. In most cases, we observed U-modal distributions with peaks at both ends of the possible values.

We found that 92.4% of the ratings of the "search goods" in the sample have a bimodal distribution, of which 88.6% have a U-modal distribution and 84.1% can be described as S-modal (and 60.5% as strictly S-modal). Conversely, we found that only 18.8% of experience goods' ratings have bimodal distributions and the majority follow a unimodal distribution. The complete analysis of the distributions of product categories is shown in Table 1.

From a quantitative point of view, we described the distribution of frequency towards an *ad hoc* distribution function. Applying a tensor-based methodology, we estimated the related value of the parameters. To evaluate the reliability of the new distribution we calculated the correlation index between the values observed in the different categories of product and the expected values of the function. We also compared these correlations with those obtained using polynomial regression. Both P_a and P_{ab} distributions' correlation index are equal to or higher than the index obtained with polynomial functions of grades two and three, and these results are reported in Table 2. The main findings are illustrated in Fig. 2.

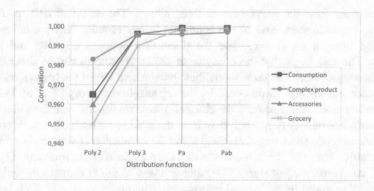

Fig. 2. Correlation index between calculated and observed data

We further identified the range of *a* parameter, as describe in the function (2) to describe the type of distribution highlighted in Table 1. We found that:

(a) for $1 > a > 0.71$ the distribution is unimodal;
(b) for $0.71 > a > 0.001$ the distribution is U-modal;
    i. where for $0.60 > a > 0.51$ the distribution is S-modal;
    ii. where for $0.71 > a > 0.60$ the distribution is strictly S-modal.

The values of the parameter were chosen to observe change in the decimals of the percentage according to the format of the collected data. Combining qualitative results with quantitative results, we found that the P distribution properly described the type of distribution among the product categories, also correctly highlighting the prevalence of strictly S-modal distributions in the categories consumption, accessories and grocery (as the estimation of the parameter *a* is consistent with the range described above). In addition, we found that the value of the parameter *a* could be roughly approximated by observing the frequency of the highest rating. We used an Amazon constant value (for our data, *k* is 58.1%) in the form of a point of accumulation, described as follows:

$$a \cong \frac{x_5 + k}{2} \tag{5}$$

This estimation was also confirmed for the "experience" category of products. At this stage, we cannot determine if (5) is the result of the e-commerce web site algorithm or if it can be generally applied to all the products, i.e., applied to the behaviors of online consumers. At this point of the analysis, we investigated the reasons for the particular distributional non-unimodality observed.

4   Discussion

Gathered data show the prevalence of three ratings. The so-called EVL preferences represent approximately 90% of the behaviors in our sample. This preliminary consideration brought us to overcome the distinction between negative and positive reviews. The data show that the behavior of online consumers can be described in a three-element schema. The three modalities individuated reflect the behavioral schema "exit, voice and loyalty" as proposed by the organizational theory [27, 28]. Following this interpretation, clients activate the channel of communication with the seller and the consumer's community only when they have the need to transmit a given information: their dissatisfaction and a negative rating about the transaction (1 star), their satisfaction and a positive rating (5 stars), a critique of a characteristic of the transaction (product, service, assistance, etc.) (4 stars). The latter is a way to say "I cannot rate with 5 stars because…" followed by the critical information. The predominance of EVL behaviors can be observed both in the reviews of the products of "search" type and in "experience" type with the same percentage. "Search" and "experience" products are different from the point of view of the large size of the frequency range. "Experience" products register a larger range, especially for the frequency of the intermediate ratings. This is principally due to the difference of the trend of distributions – "experience" type products are characterized by distributions that are mainly unimodal and monotone, i.e., their ratings have decreasing frequencies. "Search" category products are also distinguished by distributions that are polarized at the two extremes, or U-modal. Among these distributions, the percentage of tri-modal trends, or S-modal, are the majority. This percentage can be associated with the complexity of evaluation from the user's perspective and can explain the low percentage of S-modal distribution among grocery products in "search" category, as these products overlap both "search" and "experience" categories. This interpretation is also confirmed by the lowest percentage of S-modal distributions among the products in the "experience" group. A quantitative approach confirms the evidence of the qualitative analysis. Moreover, the quantitative analysis suggested a predominant role of the percentage of the maximum rating (5 stars) in the determination of the type of the distribution.

5 Conclusions

The analysis of a sample of reviews showed us that "search" category products are mainly characterized by S-modal distributions, a particular type of J-shaped distribution already identified in the literature [15]. As we study the specific action of writing a review, our analysis is different from the [15] (the author was searching for a proxy between online and real evaluation of products, normally distributed, as experimentally verified). The aim of our research is limited and the comparison with potential experimental data [29] constitutes a valid path for further analysis to evaluate the role of personal and socio-demographic characteristics, as in [30] and [1]. Though they are usually divided into negative and positive, the observed tri-modal feature of ratings allowed us to classify them using a new scheme that recalls the interpretation of the theory of organizations "Exit, Voice and Loyalty" (EVL) by [27, 28]. Given these preliminary highlights, the relation between reviews and EVL behaviors could be investigated more broadly in the future through in-depth content analysis. We also observed that the dependence of the distribution on the frequency of the maximum rating, as data suggest, is consistent with the persistent temporal dynamics individuated by [17]. As our research is focused on static ratings, this finding should be further analyzed. To enforce the capability of analysis of the observed data, we introduced an ad hoc function of distribution based on a beta-type distribution. This function presented a correlation between calculated and observed data, equal or higher than the correlation obtained with a polynomial regression using only one estimated parameter (and two estimated parameters in its early definition). In the future, the possibility of describing the distributions of ratings through a function could enable more efficient way to study larger and more heterogeneous samples, applying the methodology used in our study as a guide.

Appendix

The analysis presented in this paper is based on products sampling method. The minimum number of ratings per product is 100 for each category, while a maximum number is not configured. Therefore, the amounts of ratings vary depending on each product. We compare their distributions based on vectors of ratings with different magnitude, synthesized using percentage composition of ratings. Nonetheless, we show absolute values as well as additional information on data collected in Table 3. We noted that categories with the greater percentage of unimodal distribution of ratings, such as Clothes and Grocery, have also the lowest number of ratings collected. This finding might highlight the bounded familiarity of online users with these types of categories, suggesting a relationship between familiarity and ratings distribution. This kind of relationship will be further analyzed in future studies by using extra-lab experiments and advanced computational techniques.

Table 3. Distribution of ratings and details for categories

Type	Products	Ratings	Average ratings per product	Maximum ratings per product[a]	Distribution of ratings				
					EVL	Unimodal	Bimodal	U-modal	S-modal (Strictly S)
Consumption	333	198,836	597	9,363	310	0	333	333	333 (293)
Complex pr.	333	85,568	257	3,156	291	0	333	323	323 (161)
Accessories	333	121,655	365	5,698	294	40	293	272	272 (182)
Grocery	333	45,896	138	1,025	292	61	272	252	192 (171)
Search pr.	1132	452,045	339	9,363	1,187	101	1,231	1180	1120 (807)
Clothes/exp. pr.	333	51,192	154	324	287	212	121	121	110 (101)
All products	1665	503,237	302	9,363	1,480	313	1,352	1,302	1,230 (907)

[a]The minimum number of ratings per product is 100 for each category.

References

1. Dellarocas, C.: The digitization of word-of-mouth: promise and challenges of online reputation mechanisms. Manag. Sci. **49**(10), 1407–1424 (2003)
2. Richard, M.O., Chebat, J.C., Yang, Z., Putrevu, S.: A proposed model of online consumer behavior: assessing the role of gender. J. Bus. Res. **63**(9), 926–934 (2010)
3. Kucuk, S.U., Krishnamurthy, S.: An analysis of consumer power on the internet. Technovation **27**(1–2), 47–56 (2007)
4. Puto, C.: The framing of buying decisions. J. Consum. Res. **14**, 301–315 (1987)
5. Everhart, D.E., Shucard, J.L., Quatrin, T., Shucard, D.W.: Sex-related differences in event-related potentials, face recognition, and facial affect processing in prepubertal children. Neuropsychology **15**, 329–341 (2001)
6. Gorman, C., Nash, M.J., Ehrenreich, B.: Sizing up the sexes. Time **139**, 42–49 (1992)
7. Saucier, D.M., Elias, L.J.: Lateral and sex differences in manual gesture during conversation. Laterality **6**, 239–245 (2001)

8. Forman, C., Ghose, A., Wiesenfeld, B.: Examining the relationship between reviews and sales: the role of reviewer identity disclosure in electronic markets. Inf. Syst. Res. **19**(3), 291–313 (2008)
9. Wu, J., Gaytán, E.: The role of online seller reviews and product price on buyers' willingness-to-pay: a risk perspective. Eur. J. Inf. Syst. **22**(4), 416–433 (2013)
10. Berger, J., Sorensen, A.T., Rasmussen, S.J.: Positive effects of negative publicity: when negative reviews increase sales. Mark. Sci. **29**(5), 815–827 (2010)
11. Coker, B.L.: Seeking the opinions of others online: evidence of evaluation overshoot. J. Econ. Psychol. **33**(6), 1033–1042 (2012)
12. Wu, J.: Review popularity and review helpfulness: a model for user review effectiveness. Decis. Support Syst. **97**, 92–103 (2017)
13. Chua, A.Y., Banerjee, S.: Analyzing review efficacy on Amazon. com: does the rich grow richer? Comput. Hum. Behav. **75**, 501–509 (2017)
14. Mudambi, M., Schuff, D.: What makes a helpful online review? A study of customer reviews on Amazon.com. MIS Q. **34**(1), 185–200 (2010)
15. Hu, N., Zhang, J., Pavlou, P.A.: Overcoming the J-shaped distribution of product reviews. Commun. ACM **52**, 144–147 (2009)
16. Mavrodiev, P., Tessone, C.J., Schweitzer, F.: Quantifying the effects of social influence. Sci. Rep. **3**, 1360 (2013)
17. Sun, M.: How does the variance of product ratings matter? Manag. Sci. **58**(4), 696–707 (2012)
18. Bucciarelli, E., Persico, T.E.: Processing and analysing experimental data using a tensor-based method: evidence from an ultimatum game study. In: Decision Economics, vol. 618, p. 122. Springer (2017)
19. Wan, Y.: The Matthew effect in social commerce. Electr. Mark. **25**(4), 313–324 (2015)
20. Kousha, K., Thelwall, M.: Can Amazon.com reviews help to assess the wider impacts of books? J. Assoc. Inf. Sci. Technol. **67**(3), 566–581 (2016)
21. Feng, S., Xing, L., Gogar, A., Choi, Y.: Distributional footprints of deceptive product reviews. In: Proceedings of the Sixth International AAAI Conference on Weblogs and Social Media (2012)
22. Chevalier, J.A., Mayzlin, D.: The Effect of Word of Mouth on Sales: Online Book Reviews. NBER Working Papers 10148 National Bureau of Economic Research, Inc. (2003)
23. Nelson, P.: Information and consumer behavior. J. Polit. Econ. **78**(2), 311–329 (1970)
24. Willemsen, L.M., Neijens, P.C., Bronner, F., de Ridder, J.A.: 'Highly recommended!' the content characteristics and perceived usefulness of online consumer reviews. J. Comput. Mediat. Commun. **17**(1), 19–38 (2011)
25. Huang, P., Lurie, N., Mitra, S.: Searching for experience on the web: an empirical examination of consumer behavior for search and experience goods. J. Mark. **70**(3), 74–89 (2009)
26. David, S., Pinch, T.J.: Six Degrees of Reputation: The Use and Abuse of Online. Review and Recommendation Systems (2005)
27. March, J.G., Simon, H.A.: Organizations. Blackwell, Oxford (1958)
28. Hirschman, A.O.: Exit, Voice, and Loyalty: Responses to Decline in Firms, Organizations, and States, vol. 25. Harvard University Press, Cambridge (1970)
29. Maute, M.F., Forrester Jr., W.R.: The structure and determinants of consumer compliant intentions and behavior. J. Econ. Psychol. **14**, 219–247 (1993)
30. Ross, M.H.: Political organization and political participation: exit, voice and loyalty in preindustrial societies. Comp. Polit. **21**, 73–89 (1988)

Effective Land-Use and Public Regional Planning in the Mining Industry: The Case of Abruzzo

Francesco De Luca[1], Stefania Fensore[2(✉)], and Enrica Meschieri[1]

[1] DEA, University "G. d'Annunzio" of Chieti-Pescara,
Viale Pindaro 42, 65127 Pescara, Italy
[2] DSFPEQ, University "G. d'Annunzio" of Chieti-Pescara,
Viale Pindaro 42, 65127 Pescara, Italy
stefania.fensore@unich.it

Abstract. Land use patterns are the visible consequences of human intervention on the natural environment, especially with reference to mining activities for direct trade or subsequent manufacturing purposes. But the extractive industry and the activities of its supply chain have divergent interests related to the quantities of materials that can be extracted in compliance with the fundamental parameters of environmental sustainability. This is the main reason why mining activities are generally allowed only if public authorities issue the specific permits consistently with a specific long range plan. As the Italian legislation attributes this planning authority to regional governments, the present work aims to help to describe the socio-economic variables most affected by quarry extraction processes by referring to the case of Abruzzo (region of Central Italy). We also introduce a model for quantification of the quarry material requirements expressed by the economic operators of the same territory with a time horizon of 2020. To this end, we suggest the use of economic and statistical indicators, such as public investment on infrastructures, GDP growth, social housing policies, private building permits, to optimize the predicting power of the model as they represent reliable proxies of the demand of raw materials, in respect to the need to limit the impact on the environment.

Keywords: Land use · Public planning · Mining activities
Sustainable indicators

1 Introduction

Land use patterns are the visible consequences of human intervention on the natural environment, see [8]. The interaction between human beings and natural resources includes residential, commercial, industrial and agricultural uses, see [4]. It follows that land is shaped as a result of human policies and economic activities. Among several ways of using natural resources such as converting forests and grasslands into agricultural and grazing areas for crop and livestock production, to urban and industrial land, and to infrastructure, land is also subjected to mining activities addressed to obtain mineral resources for direct trade or subsequent manufacturing purposes.

E. Bucciarelli et al. (Eds.): DCAI 2018, AISC 805, pp. 154–161, 2019.
https://doi.org/10.1007/978-3-319-99698-1_17

In Italy, quarrying activities are defined in the original Royal Decree RD no. 1443 of 29/07/1927, and represent a significant branch of the national economy. While this legislation has been initially inspired by the privilege of the interests of production, however, the quarry materials (peat, building, road and hydraulic materials, coloring earth, fossil flour, quartz and silica sand, molar stones, stone, other industrially usable materials, not included in the first category) constitute natural resources that cannot be reproduced. Furthermore, the extraction activity interferes with the development of the territory and/or the quality of life of the residents. This means that only a rational employment of them could represent the starting point of a sustainable development of society, see [11].

With a view to the strategic management of the extraction-based economy, we must not focus solely on the obvious aspect of the consumption of non-renewable resources, and, indirectly, on the reduction of useful land. Relevant impacts rely also on the changes induced in the landscape and the possible hydrogeological and hydrographic alterations. Finally, it is necessary to consider the possibility of failure phenomena related to geomorphological modifications due to excavations.

From the point of view of environmental pollution, we concur that the mining activity generates degradation phenomena related to waste management, noise, and dust production. Finally, environmental problems also arise for activities related to the processing of the extracted materials. Therefore, the extractive industry and the activities of the relative supply chain have divergent interests related to the quantities of materials that can be extracted in compliance with some fundamental parameters of environmental sustainability.

The Italian legislation attributes the competence to the Regions about the issue of authorizations to the mining operators. Therefore, the aim of the present contribution is twofold: firstly, it aims to help to describe the socio-economic variables most affected by quarry extraction processes by referring to the case of Abruzzo (region of Central Italy). Secondly, it introduces a model to quantify the quarry material requirements expressed by the economic operators of the Region considered with a time horizon of 2020. To this end, we assume that the quantification of these future requirements is subject to numerous random factors related to the horizons of the regional plan, which are settled by the law as very long, to the variability of certain demand sectors, for example public works, and, in particular for the size of the provinces, the need to take into account the evolution of demand and extra-regional supply. A further factor of uncertainty for the forecasts is the unprecedented global economic crisis that has been going on since 2008, and whose effects and recovery times are still difficult to predict, even in the vast temporal horizon taken into consideration in the plan.

In terms of available dataset we cannot use data from the National Institute of Statistics (ISTAT) because ISTAT gathered data from questionnaires to operators who are probably not completely available to disclose all the information they have. Moreover, this kind of data is updated on a ten-year basis, and this means that current available data are too old (last update dates back to 2011).

To this research purposes, therefore, we decided to use different data sources rather than ISTAT. However, when we compared our findings with ISTAT trends of the period 1991–2011, we found a confirmation of our predicted trends.

The remainder of the work is as follows. In Sect. 2 we provide a review of the literature and a description of the socio-economic aspects of the extracting industry in Abruzzo. Then, in Sect. 3, we examine the main variables that could exert and influence in the future orientation of raw material demand. Next, we build a statistical model to assess the expected demand of materials from mining activities. This model is useful in addressing the regional policies in issuing or renewing permits to extracting operators in respect to the minimum use of land.

2 Literature Background and Extracting Industry in Abruzzo

In general, it is unanimously agreed that only a rational employment of natural resources could represent the starting point of a sustainable development of society. In fact, the use of natural resources has already attracted the attention on numerous scholars who have dealt with several connected topics.

The study presented by [8] proposes a very useful framework of the main contributions to the topic of land use and the trade-off between economics development and environmental sustainability. Particularly, authors posit that the use of land involves both land cover (such as agriculture and recreation) and actual land use, see [9]: the latter goes beyond the outward appearance where the conversion of land may impact soil, water, climate, and atmosphere (see [11]), thus resulting in affecting global environmental issues.

It follows that policy makers at a global level have more and more focused their actions on addressing land-use developments through a wide range of interventions such as limiting specific economic activities on some areas or favoring them within restricted boundaries. In doing so, policy makers, both at a global and a local level try to adopt specific models to simulate development scenarios within which support or limit economic activities, see [13].

Land-use models are used to indicate possible future land-use patterns according to the specified scenario conditions, as is demonstrated in numerous applications (see [3, 5, 12, 13]). In designing future scenarios and possible alternative solutions, policy makers have to consider optimization techniques, and calculate an optimal land-use configuration based on a set of prior conditions, criteria and decision variables, see [1] for details. To this end, mathematic and statistic programming techniques could be useful to compose different, divergent objectives of involved subjects and help policy makers in finding an optimal solution, see [10].

The mining activity is one of the more impacting activity on the land use, and several scholars have contributed to the debate from various points of view. For example, [6] have observed this phenomenon from the corporate perspective. They affirm:

"Since mining processes have the potential to impact a diverse group of environmental entities, and are of interest to a wide range of stakeholder groups, there is ample opportunity for the industry to operate more sustainably. Specifically, with improved planning, implementation of sound environmental management tools and cleaner technologies, extended social responsibility to stakeholder groups, the formation of sustainability partnerships, and improved training, a mine can improve performance in both the environmental and socioeconomic arenas, and thus contribute enormously to sustainable development at the mine level" Hilson and Murck [6] p. 227.

On the other side, [2] tries to develop a framework for sustainable development indicators for the mining and minerals industry. Particularly, this author provides specific sustainability indicators as a tool for performance assessment and improvements across different sub-industry of mining activities (i.e.: metallic, construction and industrial minerals, energy minerals, coal). The proposed indicators include economic, environmental, social and integrated variables to be used also for reporting purposes to the stakeholders collectivity.

Moreover, with a perspective to the disclosure of mining operators, [7] contribute to the debate by exploring trends in social and environmental disclosure as proxies of their corporate social responsibility. Authors affirm that Corporate Social Responsibility affect mining industry as well, where the need for individual companies to justify their existence and document their performance through the disclosure of social and environmental information is progressively increasing. Therefore, they offer a detailed review of the development of the media of social and environmental disclosure in the mining industry, and of the factors that drive the development of such disclosure.

In Italy, the extracting activity is subjected to a permit and extracting companies have to apply for obtaining it. Therefore, the public authority has to plan in advance how it intends to configure the use of land and to what extent release permits through periodic call for application from interested companies. The Italian law settles that this competence belongs to Regions that have to draft a specific plan and update it periodically. This plan should provide statistical and scientific support to the public authority in order to act consistently with a planned scenario in which interests of economic development, environmental protection and social cohesion should be composed and optimized.

As above mentioned, this work aims to provide Abruzzo with a contribution to help in planning the mining activities. This contribution serves as a case study of how the sustainable development is pursued through a rational measure of the minerals demand and the consequent release of permits in order to satisfy demand and supply. In this sense, a public policy maker is expected to achieve two goals at the same time: (i) to protect the environment and (ii) to support the economic activity.

3 Economic and Statistical Indicators for Mining Industry Planning: The Case of Abruzzo

In order to build a robust model to assess the expected demand of raw material from extracting activities, the planning process starts from the definition of specific variables able to predict the expected consumption of those materials.

To this end, we consider for the statistical analysis two kinds of variables. The first group include the "response variables", i.e. the variables that we want to predict in terms of amount of demand. Specifically, they are: *inert*, *limestone*, *chalk*, and *clay*. The second group include the "proxy variables", i.e. the variables selected to design the economic and market context which the extracting companies are operating within.

The variables that belong to this group are the *market of bricks*, the *market of concrete and cement*, the *added value of extracting industry*, and the *added value of building industry*. They basically are useful to study and understand the potential evolution path of economic context and the potential trend of demand. Historical data is available until 2014, while expected data refers to the period 2015–2020. Since the time series available concerning the quarry materials are not long enough to get accurate predictions, we start the statistical analysis by using the proxy variables.

The statistical analysis of the above variables has resulted in the trend predictions reported in Figs. 1, 2, 3 and 4

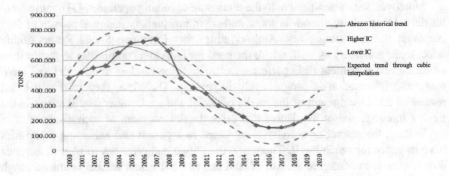

Fig. 1. The historical and expected trend of bricks market.

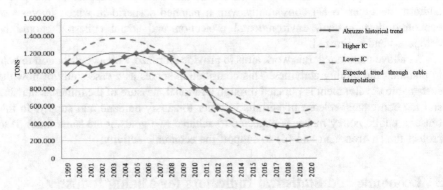

Fig. 2. The historical and expected trend of cement/concrete production.

We noticed a similarity between short-term trends in the brick and concrete quantity produced and the added value of the extractive industry. On the other side, the historical value series of the extractive industry has the advantage of being very long, and this allows for reliable long-term cycle-trend estimates.

The cycle-trend component was obtained by cubic (bricks, concrete production, added value of extracting industry) and linear (building industry) interpolation of the data. The cubic interpolation, made technically sensible by the availability of a

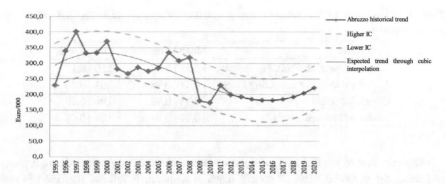

Fig. 3. The historical and expected trend of added value of extracting industry.

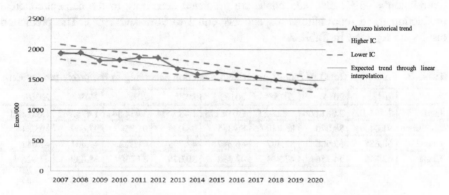

Fig. 4. The historical and expected trend of building industry.

sufficient number of data, highlights the two economic cycles that occurred during the period in question in addition to the underlying trend. In the previous figures, in addition to the point estimation, the relative 95% confidence intervals (IC) are also shown.

The interpolation analysis resulted in four possible scenarios and we propose a simple prediction tool to support the decision process of the Region in its planning activity.

We posit that a forecast of the needs of the quarry material (inert, limestone, chalk, and clay) can be based according to above forecasts. Specifically, we decided to consider the most reliable proxy variable to predict a specific type of material.

As an index of reliability we adopted the correlation coefficient between the series of quantities produced (proxy variables) and the series of values of the response variables recorded in the same years. By considering the period 2007–2012 we obtained the results collected in Table 1.

Table 1. Correlation matrix between the production of quarry materials and the related production sectors.

Correlation	Inert	Limestone	Clay	Chalk
Added value of the building industry	0,31	0,88	0,50	0,11
Added value of the extracting industry	0,47	0,95	0,64	0,26
Cement/concrete	0,93	0,88	0,96	0,79
Bricks and related goods	0,84	0,94	0,89	0,68

The selection of the highest correlation scores makes easier to adopt the best proxy to predict the potential trend of quarry material demand. It follows that inert is predicted according to the cement/concrete production trend (correlation = 93%); limestone is predicted according to the added value of extracting industry trend (correlation = 95%); clay and chalk are predicted according to the cement/concrete production trend (correlations respectively equal to 96% and 79%). The prediction results are reported in Table 2.

Table 2. Prediction (tons) of quarry materials production according to the proxy productions.

	2013	2014	2015	2016	2017	2018	2019	2020
Inert	2.384.166	2.081.600	1.822.237	1.619.175	1.485.511	1.434.341	1.478.762	1.631.871
Limestone	91.043	88.070	86.449	86.385	88.086	91.759	97.609	105.844
Clay	68.685	59.968	52.497	46.647	42.796	41.322	42.601	47.012
Chalk	62.715	54.756	47.934	42.592	39.076	37.730	38.899	42.926

4   Concluding Remarks

The above analysis has represented an example of the use of statistical and economic indicators to support the decision process. In this case, we tried to develop a reliable prediction model to help to plan the regional policy of authorizing the extraction of quarry materials, analyzing the case of Abruzzo. We noticed how the economic contingent situations, the financial crisis, the fiscal public policy are determinants of the future trend of quarry material consumption. This means that policy makers should be aware that the financial crisis starting in 2008 is still impacting on economic growth and the economic upturn is not clearly visible yet, at least in Abruzzo.

The predicting power of the developed model relies on the reliability of the selected variables in order to forecast the quarry materials demand from the markets. In this case, we developed a reliable model because of the very high level of correlation between the historical series of specific materials production (inert, clay, limestone, chalk) and the economic trend of extracting and building industry, and of the production of cement/concrete, and bricks and equivalent.

This study contributes to the debate about the sustainable economic development and the policy about environment protection. It provides a simple and effective tool to forecast the need of quarry material and accordingly decide about the release to quarry.

It is a useful tool for policy makers in their planning activity and decision process about how allowing the economic development of a territory and limiting the environmental resource consumption.

References

1. Aerts, J.: Spatial decision support for resource allocation. PhD Dissertation, Universiteit van Amsterdam (2002)
2. Azapagic, A.: Developing a framework for sustainable development indicators for the mining and minerals industry. J. Clean. Prod. **12**, 639–662 (2004)
3. De Nijs, T.C.M., de Niet, R., Crommentuijn, L.: Constructing land-use maps of the Netherlands in 2030. J. Environ. Manag. **72**(1–2), 35–42 (2004)
4. Ekabten, E.B., Mulongo, L.S., Isaboke, P.O.: Effective land-use strategies for the restoration of Maragoli forest in Vihiga county, Kenya. J. Sci. Res. Stud. **4**(11), 293–303 (2017)
5. Frenkel, A.: The potential effect of national growth management policy on urban sprawl and the depletion of open spaces and farmland. Land Use Policy **21**(4), 357–369 (2004)
6. Hilson, G., Murck, B.: Sustainable development in the mining industry: clarifying the corporate perspective. Resour. Policy **26**, 227–238 (2000)
7. Jenkins, H., Yakovleva, N.: Corporate social responsibility in the mining industry: exploring trends in social and environmental disclosure. J. Clean. Prod. **14**, 271–284 (2006)
8. Koomen, E., Rietveld, P., de Nijs, T.: Modelling land-use change for spatial planning support. Ann. Reg. Sci. **42**(1), 1–10 (2008). https://doi.org/10.1007/s00168-007-0155-1
9. Lambin, E.F., Turner, B.L., Geist, H.J., Agbola, S.B., Angelsen, A., Bruce, J.W., Coomes, O.T., Dirzo, R., Fischer, G., Folke, C., George, P.S., Homewood, K., Imbernon, J., Leemans, R., Li, X., Moran, E.F., Mortimore, M., Ramakrishnan, P.S., Richards, J.F., Skanes, H., Stone, G.D., Svedin, U., Veldkamp, T.A., Vogel, C., Xu, J.: The causes of land-use and land-cover change, moving beyond the myths. Glob. Environ. Change **11**(4), 5–13 (2001)
10. Loonen, W., Heuberger, P., Kuijpers-Linde, M.: Spatial optimisation in land-use allocation problems. Chapter 9. In: Koomen, E., Stillwell, J., Bakema, A., Scholten, H.J. (eds.) Modelling Land-use Change; Progress and Applications, pp. 147–165. Springer, Dordrecht (2007)
11. Meyer, W.B., Turner II, B.L.: Changes in Land Use and Land Cover: A Global Perspective. Cambridge University Press, Cambridge (1994)
12. Solecki, W.D., Oliveri, C.: Downscaling climate change scenarios in an urban land use change model. J. Environ. Manag. **72**, 105–115 (2004)
13. Verburg, P.H., Schot, P.P., Dijst, M.J., Veldkamp, A.: Land use change modelling: current practice and research priorities. GeoJournal **61**, 309–324 (2004)

Do ICTs Matter for Italy?

Daniela Cialfi$^{(\boxtimes)}$ and Emiliano Colantonio

Department of Philosophical, Pedagogical and Economic-Quantitative Sciences,
University G.D'Annunzio Chieti-Pescara, Viale Pindaro 42, 65127 Pescara, Italy
{daniela.cialfi,emiliano.colantonio}@unich.it

Abstract. Information and Communication Technologies (ICTs) impact the communities in which we live and the way individuals, business, government and civil society interact and develop. Simultaneously, all sectors have shown increased interest in the concept of social capital and the role it could play in building stronger communities, increasing economic productivity and contributing to regional rejuvenation. Thus, ICTs and social capital concept interlink with each other in a variety of ways. The purpose of this paper is to investigate the relationship between ICTs and social capital through the study of the relative's disparities among Italian regions. This paper provides an operational definition of the concepts of ICT and social capital and presents consistent evidence on the geography of this relationship in Italy. The statistical and geographical analysis, based on non-linear clustering with self-organizing map (SOM) neural networks, are performed to analyse the performance of Italian regions in the period 2006–2013. The results show the isolation of Southern Italian regions. In particular, we found that ICTs may not promote social capital, that is, ICTs could not play a decisive role in creating and developing social capital. These results prompt the formulation of new policies for Italian regions.

Keywords: ICT · Neural networks · Regional performance

1 Introduction

Information and Communication Technologies (ICTs) and social capital are themselves two concepts interlinked with each other in a wide variety of ways. This relationship, also, could be more ambivalent: ICTs are sometimes expected to pose challenges to the social capital in local communities, but also ICTs represent new and open opportunities for weaving new social ties and expanding the formation of social capital. Exploring this relationship is essential for at least two reasons: lots of empirical evidence indicate that social capital plays a beneficial role in health, education, public participation and the realisation of economic opportunities. Secondly, a new generation of digital ICTs, such as the Internet, have been used for better understand their impact on society, including issues about social capital.

This paper aims to map out the relationship between ICTs and social capital through the study of the relative disparities among Italian regions. The analytical tool employed for the analysis is the Self-Organizing Map (SOM), an unsupervised computational neural network. This tool is helpful for highlighting, from a topological

© Springer Nature Switzerland AG 2019
E. Bucciarelli et al. (Eds.): DCAI 2018, AISC 805, pp. 162–170, 2019.
https://doi.org/10.1007/978-3-319-99698-1_18

perspective, the evolution of the database used and for better understanding if ICTs could play a determinant role in building social capital in Italy.

This paper is structured as follow. The second section provides a literature review about the relationship between ICT and social capital. The third section regards the research methodology. It reports the variables used for the construction of the database evaluated in the analysis. The following part details the measures and the results of the Self-Organizing Map. The study ends with a part where we present the implications regarding the findings.

2   The Potential Relationship Between ICT and Social Capital: A Literature Review

Reflecting the diversity of disciplines and researchers' contributions to the debate on social capital, for the scope of the paper it was adopted the following working definition: *social capital refers to the extent, nature and quality of social ties that individuals or communities can mobilise in conducting their affairs.*

In this broad formulation, social capital encompasses a wide variety of connections and networks that people maintain with family, friends, neighbours, colleagues and at the same time, it relates to the strength of social norms, such as trust, sense of commitment and reciprocity or shared understanding that can underpin these ties. In this sense, researchers typically make a distinction, particularly crucial for this study, between:

- Bonding social capital: strong ties with the most immediate family members, closest friends and within communities of like-minded people that are bound together by shared features, considered as fundamentals of their identity;
- Bridging social capital: less committal connections to acquaintances, colleagues and weaker ties between diverse communities;
- Linking social capital: vertical interconnections between different levels of social aggregates.

In this theoretical framework, in which way ICTs interact with social capital. A firm answer to this question it is not possible, but some more general dynamics and patterns could be identified. Recent empirical evidence, like [3], have found that ICTs could enable individuals to thicken existing ties and generate new ones. ICTs in the form of mobile phone or email, for example, are used to be better in touch with close friends and family members. At the same time, ICT in the form of interest-oriented online dimension groups or networking spaces come in handling to develop more new ties to like-minded people for a variety of purposes, including professional skill and career networks[1].

From these examples, it is possible to identify the existence of a positive relationship between the two concepts because ICTs are helping to expand, transform and diversify social capital providing, for example, tools for communication and collaborative information sharing or creating meeting spaces where like-minded people can gather and socialise.

---

[1] For a large scale study see [7].

3   Methodology

This empirical study investigates the dynamics and patterns of the relationship between ICTs and social capital in Italy from 2006 to 2013, using data from Rapporto Noi Italia 2016, developed by ISTAT. Methodologically, we have performed a non-linear clustering analysis with the Self-Organizing Map (SOM), one of the most important and widely used neural network architecture[2]. During the study, we have chosen the variables in the following Table 1 because we needed to cope with two main restrictions: the data availability by Italian regions during the chosen period and the choice of aggregate indicators of social capital. Regarding the first restriction, this period allowed us to evaluate the evolution of the dataset from a pre-crisis to a post-crisis scenario. Regarding the second, distinguishing among the different conceptualisations of social capital might lead to ambiguities at the operational level because it could imply the selection of different research strategy[3]. Following this approach, the choice of regional-level indicators derived from some empirical literature like [1, 2, 4, 5, 7, 8]. After choosing the variables, indicators are divided into three macro-categories linked to the Italian experience: Social capital, ICT use and Economy. These categories reflect the different but interwoven structures of the relationship between ICT and social capital at the regional level, as could be observed from Table 1 below.

Table 1. List of performance indicators from Noi Italia 2016 (years: 2006–2013).

Function	Indicator
Social capital	Social participation activities
	Unpaid work for organisations or volunteer groups
	Founded associations
ICT use	Ordered/bought over the Internet
	Accessed the Internet
	Broadband connections
Economy	GDP per capita
	Total R&D Expenditure
	Science and Technology (S&T) Graduates

The first set of indicators considered by this research concerns social capital. We have used, as social capital proxies, variables related to the associative world because, according to [8], the associational activities are an essential framework where social

---

[2] The neural network architectures were developed by [6]. For a collection of state-of-art applications to geographical analysis see [9].

[3] During this study, social capital is considered to be an attribute of networks (or societies, regions, countries, etc.).

capital could occur and grow[4]. Following this approach, the 'People aged 14 and over who during the last 12 months they have played at least one social participation activities (years 2005–2013 - percentage values)' (variable: Social participation activities), the 'People aged 14 and over who during the last 12 months have carried out unpaid work for organizations or volunteer groups (years 2005–2013 - percentage values)' (variable: Unpaid work for organizations or volunteer groups), the 'People aged 14 and over who during the last 12 months have funded associations (years 2005–2013 - percentage values)' (variables: Founded associations) are used as indicators for social capital.

The second set of indicators we used is related to the use of ICT. Information and Communication Technologies (ICTs) impact on the communities in which we live and the way individuals, business and governments and civil society interact and develop. As a consequence, nowadays the development of social capital within communities, within or outside of geographic boundaries may depend on several elements, especially the availability and practical use of broadband connectivity. All of these elements – together with the attention to issues of online trust and confidence – will determine the quality and frequency of interactions. Following this idea, the 'People who have ordered/bought over the Internet between 3 months and a year ago (years 2005–2016)' (variable: Ordered/bought over the Internet), the 'People who accessed the Internet in the last 12 months per 100 persons with the same characteristics (years 2005–2016)' (variable: Accessed the Internet) and 'Enterprises using fixed broadband connections to the Internet (years 2005–2016 - percentage values)' (variable: Broadband connections) are chosen as proxies of the ICT usage level in a Italian regions.

The last set refers to the economic performance of the Italian regions at an aggregate level. We have used this kind of variables to capture how networks, norms and social interactions are translated into an economic asset and became economic 'capital', used to achieve their economic standing. Thus, social capital is transformed from a private good into a public one that provides collective and cumulative economic benefits. Based on this approach, the 'GDP per capita (years: 2005–2015, reference year 2010)' (variable: GDP per capita), the 'Total R&D expenditure (years: 2005–2014, percentage of Gdp)' (variable: Total R&D Expenditure) and 'Science and Technology (S&T) Graduates per 1,000 inhabitants aged 20–29 years (years: 2005–2014)' (variable: Science and Technology (S&T) Graduates) are used as indicators for the economic performance level of the Italian regions.

---

[4] If we refer to these forms of social capital under functional aspects, information and trust are vital for a network since they represent most of the network's 'intangible' resources, which help the society to achieve either economic and social outcomes (like well-being and higher employment rate) or intangible outcomes (such as sense of social security). All of these aspects occur if information represents the primary resource that individuals or groups wants to achieve through the available of social connections.

4   Patterns and Dynamics of Italian Regions: From 2006 to 2013

In this empirical study, SOM approach is used to provide a short-term evaluation of the structure of this relationship. We have chosen this tool for, at least, two reasons: first, instead of multidimensional scaling approach, its input observations are not categorised *a priori* and, as a consequence, the structure is unknown. The output network can be considered as a sort of statistical space or virtual topology in which the spatial configuration is closely linked to the spatial properties of the dataset. Secondly, this tool could understandably show complex entities. This property is particularly crucial at policy support level, where the understanding of such relationships is crucial for appropriate decision-making.

For studying the relationship between ICT and social capital, we have divided the analysis into two steps: in the first step, we have considered the evolution of the Italian regions from 2006 to 2013. In the second, we have analysed the relationship more deeply through the study of the variations between ICTs and social capital variables.

Regarding the first step, the following Fig. 1 shows the location of Italian regions[5] in 2006 (a) and the analogue in 2013 (b). If we focus first on (a) and after on (b) a mass of Italian regions located in the centre of the map move down to the bottom of the SOM map.

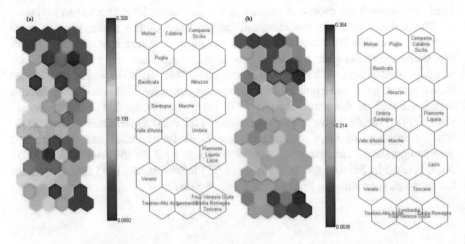

Fig. 1. Evolution of the database from 2006 to 2013.

This displacement has the effect of isolating the Southern regions from the Northern ones which remain, respectively, at the upper and bottom part of the map. As a

---

[5] The properties of the algorithm leads this map to represent the statistical property of the original dataset used to create it.

consequence, it creates a gap between them and the rest of the sample[6]. In 2013, this type of isolation caused the closeness of the Central regions to the Northern ones.

Following, we have separately analysed 2006 and 2013 through the use of their feature maps. Starting with 2006, Fig. 2 shows the feature maps for each of the nine variables that we used. The values are displayed on a gradient colour range, from blue (the lowest) to red (the highest). This tool permits:

- To obtain different profiles for the regions of the SOM;
- To discover in which socio-economic aspects the Italian regions are strong or weak;
- To understand by which means the feature maps drive the final result.

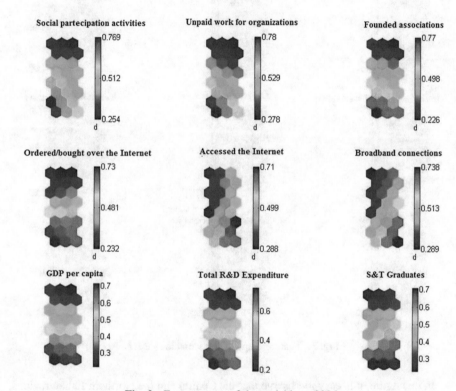

Fig. 2. Feature maps of the variables - 2006.

It is possible to observe that the Northern Italian regions, located at the bottom of the map, excels in nearly all considered variables. As profiled before into the Fig. 1, the Southern Italian regions, clustered in the upper part of the map, show low values in all

---

[6] In this context, statistical dissimilarity is translated into space distance and vice versa. This feature is obtained due to the learning nature of the algorithm. Also, it was possible because the dataset has been mapped onto the surface of the SOM by adding each neuron up to color gradient. This mapping help us to define the agglomeration of the regions. In this case, the color of the cells on the maps represents the intra cluster homogeneity degree: the cold color reveal high intra-cluster homogeneity, while the warm colors represent a lack of homogeneity.

the considered variables. In contrast, Nothern regions, such as Veneto and Trentino Alto Adige, and some Central regions, like Toscana and Emilia Romagna, obtain similar final results due to a high score in 'Social participation activities' but they poorly perform in 'Total R&D Expenditure' and 'S&T Graduates'.

Following the same kind of analysis used above, Fig. 3 shows the feature maps of the variables in 2013.

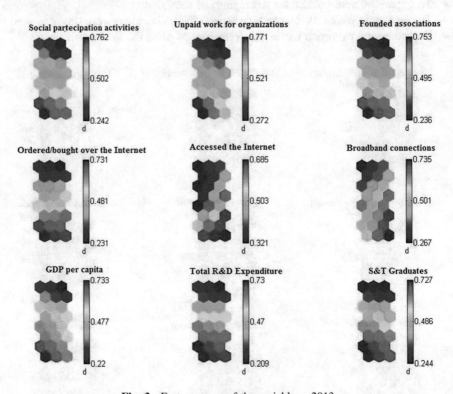

Fig. 3. Feature maps of the variables - 2013.

In this figure, it is easy to observe that the Central and the Northern Italian regions, at the bottom of the map, present high values in all the variables. As in 2006, the Southern Italian regions have low values in all the considerate variables. Instead, Northern and part of Central regions, like Liguria and Toscana; obtain similar final results due to a high level of 'S&T Graduates', 'Accessed the Internet' and ''Ordered/bough over the Internet' variables.

Focussing on the second step of the analysis, as can be deducted from the previous considerations and observing the following Table 2, from 2006 to 2013 there is a significant increase in ICT-related variables correspond to a decrease in social capital variables.

Table 2. Variations of ICT and social capital variables from 2006 to 2013.

	Social participation activities	Unpaid work for organizations or volunteer groups	Founded associations	Ordered/bought over the Internet	Accessed the Internet	Broadband connection
Piemonte	−3,4	0,3	−3,9	0,9	20,5	44,4
Valle d'Aosta-	−5,9	−2,3	−2,9	1,2	25,6	46,4
Liguria	−1,4	0	−3,6	2,9	21,6	37,1
Lombardia	−1,3	−0,4	−6,6	3	18,1	47,2
Trentino Alto Adige	−5	−1,2	−5,3	1,6	19,6	48,4
Veneto	−4,2	1,9	−3,6	2	22,5	50,8
Friuli Venezia Giulia	3,6	-0,5	−3,7	1,5	24,1	42,8
Emilia Romagna	−0,6	0,2	−6,3	1,5	21,5	47,4
Toscana	−2,5	0,6	−2,4	−1,2	20,5	46,9
Umbria	−2,7	1,1	−7,1	2,7	18,5	47,8
Marche	−3,6	2,5	−2,6	4	21,7	49,4
Lazio	0,1	1	−4,1	1	21,3	45,1
Abruzzo	0	2	−1,7	0,9	21,6	47,9
Molise	−1,8	2,8	−1,7	0,3	17,8	40,8
Campania	−3,9	0,7	−3,3	0,7	20,1	42,3
Puglia	2,8	1,7	−3,4	1,4	21,9	41
Basilicata	−2	0,3	−5,3	1,5	18,7	40,7
Calabria	−3,5	-0,2	−3,7	3,6	18,7	42,2
Sicilia	−4	0,6	−3,2	1,6	22,3	40,2
Sardegna	−6,3	1	−4,1	4,4	21,5	48

From this assumption, it emerges that in Italy ICT may not promote social capital formation, in contrast to what we have previously highlighted in the literature review. In particular, variables related to the associational activities have shown a negative relationship with ICT during the considered period of analysis. A possible explanation may be the following. All the opportunities for building and expanding social capital with ICTs are currently benefiting those are already privileged and well-endowed with social capital. Growing evidence, like in [10], suggests that only highly educated and professionally advanced people use ICT to improve their skills and networks for career advancement. As a consequence, new ICTs are unlikely to link-up people with low-networking skills and to build ties and bridges across diverse communities with different interests.

5 Conclusions

This paper has attempted to synthesise knowledge from the areas of social capital and ICT to consider if It could exist a positive relationship between ICTs and social capital in Italy.

The key findings from the case study indicate the existence of a negative relationship between the two concepts. In particular, from this negative relationship that not all social capital is positive. It could derive from the fact that the transformative impact of ICTs could be against a particular type of social capital, the social ties within local neighbourhoods.

In conclusion, the overall proposition is that through access and use of ICT communities could have more significant opportunities for engagement with others, broadening their understandings and building bonding, bridging and linking social capital. This 'whole-comminuting' perspective on the potential benefits of ICTs provides a possible focus for future researchers. For doing this, we will choose other ICTs-related variables, like the number of tweets written in Italian or the amount of Facebook accounts opened in Italy, more suitable with social capital trying to measure virtual interactions which social capital is more. Methodologically, we will improve the time-level of the analysis using a more extensive time series, moving from a short-term analysis to a long one, and after we will set-up a cause-effect analysis between the ICT and social capital variables for analysing more deeply this relationship because other external factors could interfere this relationship.

References

1. Ahmed, Z., Alzahrani, A.: Social capital and ICT intervention: a search for contextual relation. In: Proceedings of the 25th European Conference on Information Systems (ECIS), Guimarães, Portugal, pp. 2000–2016, 5–10 June 2017. http://aisel.aisnet.org/ecis2017_rp/128. ISBN 978-989-20-7655-3
2. Andriani, L., Karyampas, D.: A new proxy of Social Capital and the Economic Performance across the Italian regions. Econ. and Fin., Department of Economics, Mathematics & Statistics, University of London (2009)
3. Benkler, Y.: How Social Production Transforms Markets and Freedom. Yale University Press, New Haven (2006)
4. Degli Antoni, G.: Capitale sociale e crescita economica: verifica empirica a livello regionale e provinciale. J. Italian Econ. Ass. **3**, 363–394 (2006)
5. Gaved, M., Anderson, B.: The impact of local ICT initiatives on social capital and quality of life. Chimera Working Paper Number 2006-06 (2006)
6. Kohonen, T.: Self-Organizing Maps, 3rd edn. Springer, Berlin (2001)
7. Rainie, L., Horrigan, J.B., Wellman, B., Boase, J.: The Strength of Internet Ties. Pew Internet & American Life Project (2006)
8. Putnam, R.D.: Bowling alone: America's declining social capital. J. Democr. **1**(1), 65–78 (2000)
9. Skupin, A., Agarwal, P.: Introduction: what is a self-organizing map?. In: Agarwal, P., Skupin, A. (eds.) Self-Organizing Maps: Applications in Geographic Information Science, pp. 1–20. Wiley, Chichester (2008)
10. Zimbauer, D.: What can Social Capital and ICT do for Inclusion? Institute for Prospective Technological Studies, European Commission (2007)

Relationship of Weak Modularity and Intellectual Property Rights for Software Products

Stefan Kambiz Behfar[1]([✉]) and Qumars Behfar[2]

[1] Digital Lab, CGI Consulting, Stuttgart, Germany
stefankambiz.behfar@gmail.com
[2] Neurology Department, Cologne University, Cologne, Germany
qumars.behfar@uk-koeln.de

Abstract. There are very few studies in the literature regarding the impact of modularity on intellectual property rights, which refer to modularity of underlying products to capture value within firms. This paper brings together theory of software modularity from computer science and Intellectual Property (IP) rights from management literature in order to address the question of value appropriation for IP rights within software products. It defines the term of intellectual property associated with software products or platforms as opposed to the term of intellectual property used within particular firms serving as a source of economic rent. It initially discusses the concepts behind usage of modularity as a means to protect IP rights and explain differences of organization and product modularity, while rendering calculation for probability of imitation for weak modular systems. It investigates threat of imitation; where the main contribution of this paper is to provide a systematic analysis of value appropriation in weak modular systems by introducing a relationship between probability of imitation and module interdependency.

Keywords: Weak modularity · Intellectual property right
Probability of imitation · Module interdependency

1 Introduction

The central economic question posed by the Internet is how commercial investments can appropriate value from public good called information. Information can be transported and replicated at essentially zero marginal cost and its use by one party does not preclude use by another (Kogut and Metiu [1]). On the other hand, intellectual property (IP) as knowledge has been defined as exclusively controlled by a particular firm, which serve as a source of economic rent, such as IP patents, copyrights, and trade secrets (Carliss and Baldwin [2]). Consistent with this definition, we define IP to be a software property which excludes others from using it.

As explained by Carliss and Baldwin [2], modularity is a technical means of dividing and controlling access to knowledge and information, and can be used to

© Springer Nature Switzerland AG 2019
E. Bucciarelli et al. (Eds.): DCAI 2018, AISC 805, pp. 171–178, 2019.
https://doi.org/10.1007/978-3-319-99698-1_19

preserve IP. Modularity is a means of producing and creating new knowledge, and makes it possible to share knowledge and information about some modules, while closing off access to others, see also Simon [3, 4].

For Herbert Simon, a complex system is "one made up of a large number of parts that interact in a non-simple way. In such systems, the whole is more than the sum of the parts, at least in the important pragmatic sense that, given the properties of the parts and the laws of their interaction, it is not a trivial matter to infer the properties of the whole" [3]. Simon in another paper talked about the criterion of decomposability in modular design, which could be demonstrated both as a prescription for human designers and as a description for the systems in nature [4].

As Langlois [5] asserted, modularity is a set of rules for managing complexity. Usage of modularity in technology design is not new (Simon [3], Baldwin and Clark [6]); also Smith [7] was among the earliest stating that a modular design of social and economic institutions could make a complex modern society to be more productive. Langlois [5] investigated modularity in the design of organizations, and attempted through the literature on modular design and the property rights to create some principles for a modularity theory of the firm. Langlois also stated that "organizations reflect nonmodular structures, that is, structures in which decision rights, rights of alienation, and residual claims to income do not all reside in the same hands".

This paper makes contribution to the theoretical literature on modularity and intellectual property right. It (1) discusses the concepts behind usage of modularity as a means to protect IP rights and explains difference of weak and strong central organization platforms in terms of protecting IP rights (2) renders calculation for threat of imitation for modular systems by emphasizing that modularity is not value-neutral, but has range of weakness, which is determined based on module dependency (3) finally determines probability of imitation based on module interdependency.

2 Theory Development

We refer to the original owner of valuable information as Principal. Modularity is used to relax the principal's task of coordination by reducing interdependencies between modules. However, it requires a strong central platform which does all regulation and standardization, and let the modules do their each task, finally retrieve and assemble them. In this concept, there is a difference between organizational modularity and product/software modularity. According to Fig. 1, central platforms in organizations are usually strong, therefore the principal is less concerned if an agent defects to a competitor and violates the IP rights. In addition, it is very hard to reassemble the whole structure even if the competitor accesses to all modules.

While in software modularity, principal is more concerned if an agent defects to a competitor by providing a single module, because the "main class" in object-oriented written codes is very weak, and could be simply re-written. This means that software owner should assure that his agents do not access to all modules, otherwise if an agent defects to competitors, and whole software could be reassembled through software module interdependencies. To minimize agents' access to other modules, the modules' interdependency has to minimize, which turns weak modularity to strong. In this case, although imitation of each module could be easier due to less software module dependency and therefore less complexity, this will in turn decrease probability of whole software product imitation. According to Pil and Cohen [9], modularity decreases the difficulty of imitation for each module. Similar to software product imitation, reducing dependencies between the components mitigates complexity of each module, and could potentially increase probability of imitation for each one module via an external agent.

There is in fact no contradiction between modularity (a device to economize on control costs, mitigating asymmetries of information) and agency theory (control agents on defecting to competitors and violating the IP rights). As shown in Fig. 1, the relationship between modules and principal are strong, although principal has to reduce coordination cost. What could be weak is the relationship between modules themselves as to minimize control costs and agency costs as well as increasing value of principal.

What could be relation of a free software product and modularity (dividing and controlling access to information)? When head of development section of a software company decides to produce a new software product, one has to decide whether to sell the software product or its IP right, or alternatively put it online as a free software product and benefits from providing additional packages, services, or support. For this essential decision to make we could help by providing quantitative value for its IP right against imitation, which we claim, depends on the software weak modularity. Therefore, software is not necessarily modular structure, but in fact could be an outcome of this decision.

Last but not the least, it is important to clarify the difference between "information property" and "intellectual property", and it is significant to specify the meaning and position of those terms. Intellectual property is a common metaphor which is mostly referred by its acronym (IP). But this also restricts the way we think about information, because it has properties which make it sufficiently different from ownership of physical objects, and questions whether the property model is a good metaphor. In fact, information cannot be property; this limitation follows because information is a public good [8] suggesting that the rise of information property does not necessarily diminish the public domain because of the nature of information as an intangible public good that easily begets more information regardless of information property schemes.

A public good is a good "that can be shared non-rivalrously by many, and from whose use non-payers are not easily physically excluded. Goods with these characteristics are susceptible to free riding, and thus difficult to produce in a normal competitive market. Inventions and works of authorship are *public goods* whose creation is stimulated by the limited private exclusion rights known as patent and copyright" [9].

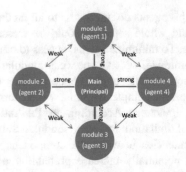

Fig. 1. Relationship between module-module and module-principal.

3 Analysis and Results

3.1 Methodology

After the last conceptual section, we discuss the methodology involving agency theory and hazard model to explain threat of imitation. Any person might be a threat to imitation of the IP product. If imitation by third parties is likely, the value of the monopoly will certainly drop. Let the flow of profit from the IP product is denoted by v, and its capitalized value in perpetuity is denoted by $V \equiv v/r$; where r is the discount rate. An agent who has access to the IP product defecting to a competitor receives the reward x, whereas its capitalized value becomes $X \equiv x/r$. According to (Carliss and Baldwin [2]) to model imitation by third parties using hazard model we assume that Ps represents the probability of imitation in any time. The probability P can take on values between zero and one, and parameter s captures all other determinants of the imitation probability [2]. We consider in case of imitation or substitution, the principal's per-period cash flow reduces to x and his establishment can worth X. Thus the principal receives surplus cash flow of v–x as long as the monopoly continues. Under these assumptions, the probability of the monopoly surviving from t to t + 1 is (1 − Ps). Using the perpetuity formula, the value of the monopoly under this threat is obtained as:

$$V = q(Ps) \cdot (v - x - NPx)$$
$$\text{where} \quad q(Ps) = (1 - Ps)/(r + Ps) \tag{1}$$

This presents that that the value of the monopoly can be subdivided into two parts: (1) excess cash flows represented by $(v - x - NPx)$ that continue as long as the monopoly continues; and (2) a capitalization factor shown by $q(P_s)$ that takes into account the probability (P_s) that the monopoly ends in any time. Obviously, a positive probability of imitation $(P_s > 0)$ can drop the value of the monopoly to the principal [2]. N is the number of agents with access to the principal's information. Cost of protecting the principal's information against imitation by agents is NX. If the value of the monopoly before agent payments, V–X, is less than NX, then it is not worth protecting. As above-mentioned, eliminating inter-dependencies between design elements in different components reduces probability of imitation for whole system.

According to Carliss and Baldwin [2], the overall impact of a given modularization should balance its effect on agent payments against the threat of imitation or substitution. This trade-off must be evaluated module-by-module, using the subscript m to refer to a particular module. The reward per person for defection in non-modular system is PX and in modular system is equal to $P X_m$. When probability of imitation is equal to Ps_m then modularization is worthwhile if the total value under modularization is greater than the monopoly value.

$$V_m(Ps) = q(Ps_m) \cdot (v_m - x_m - NPx_m) \tag{2}$$

where (Ps) is probability of imitation or substitution, and $(1 - Ps)$ is the probability of the monopoly surviving. Probability of imitation or substitution is higher in each module of the modular system than in the corresponding one-module system [2], thus $s_m > s$. When a particular modularization increases or decreases, the total value will depend on these countervailing mechanisms which operate module-by-module and aggregate to determine the sum of module values as given below:

$$\sum V_m(Ps) = \sum q(Ps_m) \cdot (v_m - x_m - NPx_m) \tag{3}$$

Probability of imitation should be correlated to module interdependency. Modular system does not necessarily mean independent components; change in one part of software might require change in other parts. However, more modularity means less module interdependency, as discussed in the last section.

3.2   Discussion

Within software products, modularity is always imperfect or weak, meaning that features (package, class, or method) always depend on each other to some extent. When a system is fully non-modular or dependent, the probability of imitation for whole system is 1, i.e. each module has access to all other modules, and if one module is imitated by an agent, the whole system, product or platform could be reassembled; whereas when the system is fully modular, the probability of imitation for whole system is 0, i.e. there is no access to all modules, because no agent has access to all the modules. Therefore in weak modular system, probability of imitation is proportional to module interdependency. This is similar to weak modularity within firms, to be discussed in the next section, where probability of imitation is proportionate to population performance which itself is a function of firms' interdependencies.

Module interdependency for software has been investigated in computer science, so-called degree distribution. It was found to be a power law $P_s \sim 1/s^\mu$, where μ ranges from 1.32 to 1.86, as predicted Potanin et al. [12], and indicated by Ma et al. [13]. Behfar et al. [14] performed a thorough study on class-class and package-package dependencies within open source software corpus network and obtained μ in a sample case equal to 1.2 and 1.6, respectively. In the next section, we attempt to verify imitation and interdependency in firms, and explain the similarities.

3.3 Verification

This section is mainly to verify the relationship between imitation and interdependency within firms. The main objective of the analysis is to test how altering design complexity affects trade-off between innovation and imitation. According to Ethiraj et al. [15], performance of firms depends on the setting of the decision variables and the interactions among them. If there are no interactions between decision variables, each decision contributes independently to overall firm performance. Alternatively with increase of interactions between decision variables, the contribution of each decision choice to firm performance becomes more interdependent. As an example, a decision choice which results in local performance gains does not necessarily lead to a concomitant increase in firm-level performance.

According to Ethiraj et al. [15], the performance contribution of each decision variable (d_i) is determined both by the state of the ith decision choice and the states of the j other decision choices on which it depends. By holding the pattern of interdependencies constant across all firms within a given scope, it results in 2N distinct fitness values, which is in fact one for each possible configuration of the N decision variables. Let R be total number of interdependencies which is distributed among N(N-1) cells in an interaction matrix. By permitting the heterogeneity in the distribution of R inter-dependencies across all firms, the total number of distinct performance values is given by:

$$\left[2^N \binom{N(N-1)}{R} \right] \text{ where } R < N(N-1) \tag{4}$$

As explained by Ethiraj et al. [15], each firm attempts to engage in incremental innovation. Variance in the success of innovation generates heterogeneity in firm performance. The heterogeneity later fuels imitation efforts. Low performance firms engage in imitation of high performers (leaders). Imitators imitate a module from a randomly chosen leader where the leader was chosen with a probability proportionate to population performance. This is a summation of binomial distribution of R interdependencies distributed among N(N-1) cells, which indicates that the probability of imitation is proportional to interdependencies among firms. This is similar to what we have obtained in software product in probability of imitation being proportional to module interdependency.

4 Conclusion

This paper brought together theory of software modularity from computer science and intellectual property right from management literatures to address the question of value appropriation for IP right. The paper first discussed the concepts behind usage of modularity as a means to protect IP rights and explained difference between weak and strong central organization platforms in terms of protecting IP rights. However, the main contribution of this paper is to provide a systematic analysis of value appropriation in weak modular systems by determination of probability of imitation through its relationship with module inter-dependency within software products.

Value-based modular system was defined, as opposed to previously-defined value-neutral modular systems for capturing value in systems with distributed modularity (module dependency). A particular modularization increases or decreases the total value of monopoly which depends on how these countervailing mechanisms operate module-by-module and aggregate to determine the sum of monopolies as in (3).

When a system is fully non-modular or dependent, the probability of imitation for whole system is 1, since one agent from one module has access to all other modules; therefore the whole system could be easily reassembled specially in weak central platforms such as software products. Whereas when the system is fully modular, the probability of imitation for whole system is 0, because one agent has no access to all other modules. Therefore in weak modular systems, probability of imitation is proportional to module interdependency. Probability of module dependency for software has been investigated in computer science, so-called degree distribution. It is found to be a power law $P_s \sim 1/s^\mu$, where μ ranges from 1.32 to 1.86.

When head of development section of a software company decides to produce a new software product, one has to decide whether to sell the software product or its IP right, or alternatively put it online as a free software product and benefits from providing additional packages, services, or support. For this essential decision to make we could help by providing quantitative value for its IP right against hazard of imitation, which we claim, depends on the software weak modularity and interdependency.

References

1. Kogut, B., Metiu, A.: Open-source software development and distributed innovation. Oxf. Rev. Econ. Policy **17**(2), 248–264 (2001)
2. Carliss, Y., Baldwin, J.H.: The Impact of Modularity on Intellectual Property and Value Appropriation. Harvard Business School (2012)
3. Simon, H.A.: The architecture of complexity. In: Proceedings of the American Philosophical Society (1962)
4. Simon, H.A.: Near Decomposibility and Speed of Evolution. Carnegie Mellon University, Pittsburgh (2000)
5. Langlois, R.N.: Modularity in technology and organization. J. Econ. Behav. Organ. **49**, 19–37 (2002)
6. Baldwin, C.Y., Clark, K.B.: Managing in an age of modularity. Harvard Bus. Rev. **75**(5), 84–93 (1997)
7. Smith, A.: An Enquiry into the Nature and Causes of the Wealth of Nations, Glasgow edn. Clarendon Press, Oxford (1976)
8. Wagner, R.P.: Information wants to be free: intellectual property and the mythologies of control. Columbia Law Rev. **103**, 995–1034 (2003)
9. Gordon, W.J.: Authors, publishers, and public goods: trading gold for dross. Loyola Los Angeles Law Rev. **36**, 159–200 (2002)
10. Gordon, W.J.: Asymmetric market failure and prisoner's dilemma in intellectual property. 17 U. Dayton L. Rev. **853**, 854 (1992)
11. Pil, F., Cohen, S.K.: Modularity: implications for imitation, innovation, and sustained advantage. Acad. Manag. Rev. **31**(4), 995–1011 (2006)

12. Potanin, A., Noble, J., Frean, M., Biddle, R.: Scale-free geometry in OO programs. Commun. ACM **48**, 99–103 (2005)
13. Ma, J., Zend, D., Zhao, H.: Modeling the growth of complex software function dependency networks. Inf. Syst. Front. **14**(2), 301–315 (2012)
14. Behfar, S.K., Turkina, E., Cohendet, P., Burger-Helmchena, T.: Directed networks' different link formation mechanisms causing degree distribution distinction. Phys. A **462**, 479–491 (2016)
15. Ethiraj, S.K., Levinthal, D., Toy, R.R.: The dual role of modularity: innovation and imitation. Manag. Sci. **54**(5), 939–955 (2008)

Author Index

© Springer Nature Switzerland AG 2019
E. Bucciarelli et al. (Eds.): DCAI 2018, AISC 805, p. 179, 2019.
https://doi.org/10.1007/978-3-319-99698-1

Printed in the United States
By Bookmasters